S0-BVO-412

Remote Sensing
of Earth Resources

GEOGRAPHY AND TRAVEL INFORMATION GUIDE SERIES

Series Editors: Alberta Auringer Wood, Information Services Librarian, Memorial University of Newfoundland, St. Johns and Clifford H. Wood, Assistant Professor, Department of Geography, University of Newfoundland, St. Johns

Also in this series:

CARTOGRAPHY—*Edited by Harry Steward**

CULTURAL AND POPULATION GEOGRAPHY—*Edited by Robert E. Burton**

HISTORICAL GEOGRAPHY OF THE UNITED STATES AND CANADA—*Edited by Ronald E. Grim**

HISTORY OF GEOGRAPHY—*Edited by Allen B. Bushong**

TRAVEL IN ASIA—*Edited by Neal Edgar and Wendy Ma**

TRAVEL IN CANADA—*Edited by Nora Corley Murchison**

TRAVEL IN OCEANIA, AUSTRALIA, AND NEW ZEALAND—*Edited by Robert E. Burton**

TRAVEL IN THE CARIBBEAN—*Edited by Sian Steward**

TRAVEL IN THE UNITED STATES—*Edited by Joyce Post and Jeremiah Post**

URBAN GEOGRAPHY—*Edited by Nancy Hultquist and John Hultquist**

*in preparation

The above series is part of the
GALE INFORMATION GUIDE LIBRARY

The Library consists of a number of separate series of guides covering major areas in the social sciences, humanities, and current affairs.

General Editor: Paul Wasserman, Professor and former Dean, School of Library and Information Services, University of Maryland

Managing Editor: Denise Allard Adzigian, Gale Research Company

Remote Sensing of Earth Resources

A GUIDE TO INFORMATION SOURCES

Volume 1 in the Geography and Travel Information Guide Series

M. Leonard Bryan

Jet Propulsion Laboratory
Pasadena, California

Gale Research Company
Book Tower, Detroit, Michigan 48226

Library of Congress Cataloging in Publication Data

Bryan, M Leonard.
    Remote sensing of earth resources.

    (Geography and travel information guide series ; v. 1)
(Gale information guide library)
    Includes indexes.
    1. Remote sensing—Bibliography. 2. Remote Sensing
—information services. I. Title II. Series
Z6004.R38B79   [G70.4] 016.621367   79-22792
ISBN 0-8103-1413-4

VITA

M. Leonard Bryan is an instructor in geography at the Glendale (California) Community College. He received his training in geography at Indiana University (A.B., 1963), McGill University (M.Sc., 1965) and the University of Michigan (Ph.D., 1971) following a four-year tour as an aerographer-meteorologist with the U.S. Navy. His research interests have centered around geomorphology, surface water hydrology and urban morphology. With respect to remote sensing, he was employed by the Environmental Research Institute of Michigan (Ann Arbor) as an assistant research geographer (1971-1975) to study the utility of radar remote sensing in environmental applications. In 1975 he transferred to the Jet Propulsion Laboratory (Pasadena) as a senior scientist involved in radar imagery interpretations. Recent research projects have included the study of sea and lake ice conditions and urban morphology based on airborne and satellite borne synthetic aperture radars and Landsat data. He is presently a division representative for the Venus Orbiting Imaging Radar Project and a member of the technical staff at the Jet Propulsion Laboratory. Past academic positions have included instructor in geography (Frostburg State College, Maryland), and lecturer in geography (the University of Michigan, Flint).

CONTENTS

Introduction   ix
Abbreviations and Acronyms   xiii

Chapter 1.   General Literature   1

Chapter 2.   Proceedings   55

Chapter 3.   Manuals and Guides   81

Chapter 4.   Catalogs   93

Chapter 5.   Maps   107

Chapter 6.   Bibliographies   113

Chapter 7.   Journals   127

Chapter 8.   Workshops, University, and Training Courses   143

Author Index   149
Title Index   157
Subject Index   167
NTIS Accession Number Index   181
Series Index   185

INTRODUCTION

The concept of remote sensing is not exceptionally new to the field of earth sciences. Indeed, aerial photography has been used at one level or another for surveillance and observation of the earth's surface and man's activities thereon for over a century. However, it is only recently that remote sensing has blossomed forth as a popular method of earth science data collection. The use of aerial photographs in the 1920s and the progressive development of film and camera techniques through the 1940s (as with the development of infrared films) are well documented and have become very standard techniques. More recently, the collection of data from aircraft, using camera systems and multiple cameras which record only selected portions of the electromagnetic spectrum, has expanded our remote-sensing horizons considerably. Adding to this, the use of instruments to detect electromagnetic radiation which is beyond the visible and near visible (e.g., microwave and ultraviolet) has yielded a broader range of data which have become available. Then, include in addition to the passive remote-sensing systems those active systems (distinguished from the former in that they provide their own illumination or energy source) which may operate in the same wavelengths of the spectrum, the field has expanded by several more times in both the data availability and complexity. Hence, where we have a camera (passive) we also have the capabilities of adding a flash unit and making the system active; the (passive) microwave radiometer is supplemented by the (active) microwave radar.

With increased interest in earth resources, and the need to collect data over extensive and often inaccessible regions as well as areas which are well populated, and with the continuing efforts of the various governments in the development and implementation of instruments to be used from space for both earth and other planetary observations, the field of remote sensing has grown considerably. Various organizations present symposia, numerous authors have prepared texts of widely varying quality, several journals have been initiated (and also terminated) to deal especially with remote sensing, and indeed a very standard American journal (PHOTOGRAMMETRIC ENGINEERING, now PHOTOGRAMMETRIC ENGINEERING AND REMOTE SENSING) changed its name in 1975 to reflect the growing interest in remote sensing. Hence, we can conclude that the field is quite active. Because of the newness, the popularity, and the amount of funds which have been injected into the area of research and applications, the philosophies, work, applications, results, and publications often seem to lack a commonality which we have come to expect in the more mature and somewhat more organized endeavors.

This guide to information sources, then, is an effort to scan the literature of the remote sensing of earth resources. It is divided into eight major sections: general literature, proceedings of symposia and meetings, manuals of remote sensing including notes from tutorial courses, catalogs of remote-sensing data available to the general public, a selection of maps which have been derived (primarily) from remotely sensed data, bibliographies focusing on remote-sensing work, major journals which often contain reports and publications of work dealing with remotely sensed data, and workshops and training courses. The eighth section, a listing of some short courses and training sessions in remote sensing, is possibly the portion of the guide which is the most active and for which it is the most difficult to maintain accuracy and timeliness. The last section consists of a set of indexes, including personal and corporate authors as well as NTIS Accession number, series, subject, and title.

With respect to the publications, the compiler has obtained each publication (a few with considerable difficulty) and, in general, they are readily available to the interested individuals. Because much of the remote-sensing work has been sponsored by various government agencies (primarily by the National Aeronautics and Space Administration, NASA) and because the reports are often the results of contractual arrangements, these reports are normally available through the National Technical Information Service for very reasonable prices and often on microfiche. Also, interestingly, because the field of remote sensing is so new, very few commercial publications in the field are now out of print thus aiding a researcher considerably in the building of his own library and resource files.

One of the major problems which has so often been apparent in bibliographies and resource guides is the lack of adequate commentary about the referenced work. Hence, there has been a concentrated effort to include some comments after each of the references in order to aid the researcher in identifying those items which may warrant pursuit. Some of these comments are the abstracts which have been gleaned from the original works, some are condensations of the introductory statements by the various authors, and the remaining have been especially prepared for this present bibliography. However, it must be recognized that these comments are massive condensations of the works and should there be a hint of interest for any particular researcher, they are strongly encouraged to check any reference to see if it would actually fulfill their purposes.

It would have been a rather simple (but obviously laborious) task to include all of the journal articles, contract reports, and other publications which fall within the field of "remote sensing of earth resources." However, this would have fulfilled little purpose for (1) the publications are changing quite rapidly and the list would have been cumbersome at best and quickly out of date, (2) one would have to search extensive lists of publications dealing with specific problems to determine what would be of particular interest, and, (3) several organizations now print complete listings of all government and open literature material on a bimonthly basis (see National Technical Information Service and American Institute of Aeronautics and Astronautics in the indexes for examples). There are several bibliographies which also summarize these works (see for example, Krumpe, 1976, item no. 299). Hence, this guide to information sources lists bibliographies and journals on the assumption that these particular references will introduce the interested student of the subject to the extensive literature which is available.

ABBREVIATIONS AND ACRONYMS

AAAS   American Association for the Advancement of Science

AGARD   Advisory Group for Aerospace Research and Development (A NATO Group)

AGI   American Geological Institute

AIAA   American Institute of Aeronautics and Astronautics

ASP   American Society of Photogrammetry

CCRS   Canada Centre for Remote Sensing, a government facility in Ottawa, Ont., Canada

CIT   California Institute of Technology, Pasadena, CA.

COSPAR   Committee on Space Research, an International Organization

EM   Electromagnetic (radiation)

EOS   Earth Observation Satellite, a generic designation for the next generation of satellites (after Landsat) *

ER   Earth Resources

ERAP   Earth Resources Aircraft Project, name applied to the NASA aircraft operations based at Moffett Field, CA.

ERS   Earth Resources System

EREP   Earth Resources Experiment Package (Skylab Project)

ERIM   Environmental Research Institute of Michigan, a non-profit research company

ERTS   Earth Resources Technology Satellite, renamed Landsat, 13JAN75 *

EROS   Earth Resources Observation System
            EROS Data Center - a major source of remotely sensed imagery and data operated by the U.S. Department of Interior (U.S. Geological Survey) in Sioux Falls, S.D.

ESA   European Space Agency

Abbreviations and Acronyms

ESMR	Electrically Scanning Microwave Radiometer, an instrument on the Nimbus satellite series
ESRO	European Space Research Organization
ESSA	Environmental Science Services Administration, a portion of the U.S. Department of Commerce
ETL	Engineer Topographic Laboratories, a facility operated by the U.S. Army Corps of Engineers located at Ft. Belvoir, VA.
GCP	Ground Control Point
GHz	Gigahertz (hertz = cycle per second)
GSFC	Goddard Space Flight Center, a NASA facility in Greenbelt, MD.
HRPI	High Resolution Pointable Imager
IAF	International Astronautical Federation
IAA	International Aerospace Abstracts, a publication of AIAA under contract from NASA
IEEE	Institute of Electrical and Electronics Engineers
IR	Infrared, a portion of the EM spectrum
ITOS	Improved TIROS Operational System
IUFRO	International Union of Forest Research Organizations
IURS	International Union of Radio Science. See also: URSI
JPL	Jet Propulsion Laboratory, a facility in Pasadena, CA., operated by the California Institute of Technology under contract from NASA
JPRS	Joint Publication Research Service
LARS	Laboratory for Applications of Remote Sensing, at Purdue University, West Lafayette, IN.
Landsat	Name applied to a series of satellites operating in the visible and near visible portion of the EM spectrum. Formerly named ERTS.*
MGI	Military Graphic Information
MSC	Manned Spacecraft Center, a NASA facility in Houston, TX., now renamed the (L.B.) Johnson Space Center.
MSS	Multispectral Scanner, a remote sensing instrument
NAS	National Academy of Science, U.S.
NASA	National Aeronautics and Space Administration, a U.S. government organization
NATO	North Atlantic Treaty Organization
NDPF	National Data Processing Facility
NESS	National Environmental Satellite Service, a portion of NOAA
NOAA	National Oceanic and Atmospheric Administration, a portion of the Department of Commerce

NRC   National Research Council, U.S.

NTIS   National Technical Information Service, Springfield, VA., a central clearinghouse for U.S. government publications in technology and science

ONC   Operational Navigation Chart, published by the U.S. Defense Mapping Agency Plan Position Indicator, a type of radar display

PPI   Plan Position Indicator

RADAR   Radio Detection And Ranging - this was originally an acronym which has become a palindrome in the English language - a remote sensing instrument

SAR   Synthetic Aperture Radar

SCS   U.S. Soil Conservation Service

Seasat   Name applied to a satellite, operating primarily in the microwave portion of the EM spectrum and designed for oceanographic research

SLAR   Side Looking Airborne Radar

SLR   Side Looking Radar

SMS   Syncnronous Meteorological Satellite

SNR   Signal to Noise Ratio

STAR   Scientific and Technical Aerospace Reports, a publication of NASA

THIR   Temperature, Humidity Infrared Radiometer

TIROS   Television Infrared Observation Satellite, an operational weather satellite. See also ITOS. *

TM   Thematic Mapper

UNESCO   United Nations Economic, Scientific and Cultural Organization

URSI   Union Radio Scientific Internationale, a group formed in 1919 for world wide cooperation in radio research. See also IURS.

USDA   U.S. Department of Agriculture

USDI   U.S. Department of Interior

USGRDR   United States Government Research and Development Reports

USGS   United States Geological Survey, a portion of the USDI

UTM   Universal Transverse Mercator, a type of map projection

UV   Ultraviolet, a portion of the EM spectrum

WHOI   Woods Hole Oceanographic Institution, a research institute located at Woods Hole, MA.

---

*   NASA satellites are given sequential letter designations prior to launch, and number designations after launch. Hence, ERTS-A was renamed ERTS-1 after launch. Similarly, NIMBUS-F was renamed NIMBUS-6 following the launch.

Chapter 1
GENERAL LITERATURE

Much of the material which is presently available with respect to remote sensing is contained not in technical reports but in general publications such as textbooks and extensive journal articles. Many of these items have been prepared by individuals working under grants and contracts from various governmental and private organizations. This portion of the information guide concerns these types of publications. Thus it contains numerous texts as well as general and descriptive publications which are oriented toward the general public interested in obtaining an overview of the field of remote sensing.

As mentioned earlier, because the field of remote sensing (with the possible exception of aerial photography) is relatively new, many of these references will be quickly superseded and replaced during the next several years. Such new publications will deal primarily with the addition of supplemental chapters to the works rather than the replacement of the existing works due to changes in the physical theory of the discipline. Consequently, the reader is advised to check for the most recent editions, especially for those references which have a textbook character.

These references, therefore, deal with overviews of remote sensing; some studies concerned with the philosophy, rationale, design, and execution of several satellites and remote-sensing instruments (e.g., SEASAT); and some theoretical and conceptual studies of systems which were never actually developed. Arguments for and against specific designs, both from the point of view of their technical feasibility as well as their scientific usefulness and import are also contained in several of the references included in this "General Literature" section.

1. Abdel-Hady, Mohamed, and El-Kassas, Ibrahim A. REMOTE SENSING RESEARCH PROJECT: GLOSSARY OF AERIAL PHOTOGRAPHY AND REMOTE SENSING IN GEOLOGY AND EARTH SCIENCES. Arab Republic of Egypt, Academy of Scientific Research and Technology. Remote Sensing Research Project, Interim Report no. 3. Stillwater, Oreg.: December 1972. 72 p. Available from NTIS, PB-220-154.

1

This glossary includes definitions of 470 terms which are commonly used by the photogeologists and the interpreters of remote sensing and aerial photography in their fields of applications in geology and earth sciences. This glossary was compiled to provide a standard set of definitions for the technical terms commonly used in aerial photography. Arranged alphabetically, it is the third interim report on the research project titled "Remote Sensing Techniques in the Arab Republic," and complements the previous reports with its explanations.

2. Alaska Remote Sensing Symposium, Anchorage, 1969. THE USE OF REMOTE SENSING IN CONSERVATION, DEVELOPMENT, AND MANAGEMENT OF THE NATURAL RESOURCES OF THE STATE OF ALASKA. Richard H. Eakins, Jr., and Robin I. Welch, co-directors. Juneau: Department of Economic Development, 1969. 133 p. Paper.

Although the publication was sponsored by the state of Alaska, many of the examples used by the authors of the twenty-five contributions dealt with little of the Alaska scene. Rather, as stated in the preface, this collection of papers "has been prepared in an effort to define the remote sensing systems most applicable to a variety of problems in Alaska." Hence, the papers concentrate on describing sensor potentials and some of the problems of Alaska to which these potentials may be applied.

3. Alexander, Larry, et al. "Remote Sensing: Environmental and Geotechnical Applications." DAMES AND MOORE ENGINEERING BULLETIN (August 1974): 1-50. Refs., illus., figs.

This issue of the DAMES AND MOORE ENGINEERING BULLETIN introduces a number of the commercially useful techniques of remote sensing, briefly describes the sensors which gather the data, and discusses some of the environmental and geotechnical areas which have benefited from the use of remote-sensing technology.

4. Ambionics, Inc. SURVEY AND ANALYSIS OF POTENTIAL USERS OF REMOTE SENSING DATA, FINAL REPORT: STATE GOVERNMENT ACTIVITIES IN REMOTE SENSING. Washington, D.C.: Ambionics, November 1975. 222 p. Bibliog. Available from NTIS, N76-23664.

Remote-sensing applications for the activities of regional interstate organizations, the federal agencies, and the private sector are examined. The survey covers activities in all fifty states. Emphasis was placed on on-going operational programs and no attempt was made to cover the activities of the federal agencies except insofar as they impinged on state or other regional or metropolitan programs. An excellent source for defining the amount, type, and direction of involvement of the states in remote-sensing activities. Includes the names and addresses of individuals and agencies contacted for this survey. A bibliography of materials, including numerous newsletters, annual reports, and similar documents from state agencies is included.

5. Anderson, Frank W., Jr. ORDERS OF MAGNITUDE: A HISTORY OF NACA AND NASA, 1915-1976. NASA SP-4403. Washington, D.C.: U.S. National Aeronautics and Space Administration, 1976. ix, 100 p. Available from U.S. Government Printing Office.

Because NASA is such an important and active participant in the remote sensing of earth resources, this concise history of NASA and its predecessor, NACA, provides some very interesting background to the development of the satellite and earth resources remote-sensing programs. The early chapters deal with various rocket programs; of special interest are the last several chapters which include outlines of the Skylab and ERTS projects. Quite interesting reading.

6. Arkell, R., and Hudlow, M. GATE INTERNATIONAL METEORO-LOGICAL RADAR ATLAS. Washington, D.C.: Center for Experimental Design and Data Analysis, 1977. iii, 222 p. Available from U.S. Government Printing Office.

During the summer of 1974 a program was conducted between the east coast of Africa and the west coast of South America. The overall project was termed the Global Atmospheric Research Program (GARP) and the subsection identified as GATE (GARP Atlantic Tropical Experiment) is the subject of this publication. This publication consists of a number of radar images of precipitation in an area centered roughly at 9°N latitude, 24°W longitude and collected between late June and late September 1974. The photomosaics provide significantly expanded aerial coverage over that obtainable from a single radar image. These are intended for use by researchers and should prove useful in tracing the morphology of echo cluster systems and for comparing these systems with the structure of cloud clusters as revealed by satellite data. The rapid changes in cloud coverage is dramatically shown in the sequences of radar mosaics.

7. Avery, T. Eugene. INTERPRETATION OF AERIAL PHOTOGRAPHS. 3d ed. Minneapolis: Burgess Publishing Co., 1977. xi, 392 p. Figs., illus., appendixes, refs.

This is a standard textbook on aerial photo interpretations, with special chapters on aerial photo orientation, geometry, stereoscopic parallax, stereograms, films and filters, sources of aerial photos, and applications in various earth sciences. Introductory statements about nonphotographic techniques and excellent examples of imagery and photography are included. Appendixes cover a glossary, course outline, and conversion tables.

8. Bailey, C.H. THE ELECTROMAGNETIC SPECTRUM AND SOUND: 50 MODERN EXPERIMENTS. New York: Pergamon Press, 1967. iv, 135 p. Figs.

The book consists of six major sections (sound, light, infrared, ultraviolet, radio, and microwaves) with each section containing numerous experiments about wave motion which can be studied with very unsophisticated equipment. Especially good for introductory statements about the physics of remote sensing. It is oriented toward high school and junior college audiences.

9. Barrett, Eric C., and Curtis, Leonard F., eds. ENVIRONMENTAL REMOTE-SENSING 2: PRACTICES AND PROBLEMS. London: Edward Arnold, 1977. 319 p. Figs., illus., refs.

Questions of policy making for remote sensing are examined in this publication, taking into account the current systems and services for remote sensing in relation to common user requirements, commercial consideration in remote-sensing engineering, national programs for remote sensing, and remote sensing from Spacelab. The processing and presentation of remote-sensing data are considered along with aspects of in situ observations and the interpretation of remote-sensing data. Attention is given to the use of remote-sensing data in: (1) cartography, (2) densitometric methods of processing remote-sensing data, (3) optical processing as an aid in analyzing remote-sensing imagery, (4) image-processing system applied to earth-resource imagery, (5) objective generalization of Landsat images, and (6) problems in analyzing and interpreting data from meteorological satellites.

10. Bay, Sally M. FINAL REPORT ON STATE USE OF SATELLITE REMOTE SENSING. Denver: National Conference of State Legislatures, 25 August, 1976. 142 p. Tables, refs., appendixes. Available from NTIS, N77-10619.

This volume details the final report and recommendations of the Task Force on Uses of Satellite Remote Sensing for State Policy Formulation of the National Conference of State Legislatures. The purpose is to review the applications and limitations of the NASA Landsat Follow-On Program which is to assist state programs related to natural resources. The following areas are reviewed: (1) the proposed capabilities of the Landsat Follow-On Program, (2) existing satellite applications

4

by state agencies, (3) existing federal and state legislation
mandating natural resources related state programs, (4) inferred data needs of those programs, and (5) an analysis of
the feasibility of the Follow-On Program to meet those needs.
A set of interesting recommendations is included, together
with four appendixes which give detailed information on
which the report is based.

11. Beatty, F.D. GEOSCIENCE POTENTIALS OF SIDE-LOOKING RADAR.
2 vols. Prepared for NASA by U.S. Army Corps of Engineers under
Contract DA-44-009-AMC-1040 (X). Alexandria, Va.: Raytheon
Autometric Corp., September 1965. Paper (spiralbound).

This report is probably one of the best overall reports thus
far presented in the unclassified literature dealing with radar
remote sensing. Unfortunately, it is out of print and copies
are extremely difficult to locate. In two volumes of large
format (12 x 14 inches), it presents both text and numerous
images obtained from a wide assortment of imaging radar
systems. Following an introduction which includes: resumés
of remote sensing, radar theory and operation, radargrammetry, image interpretation and mosaicking, approximately
fifty pages of explanations and interpretations of the images
complete the first volume. Volume 2 consists of images and
interpretation (acetate) overlays. Although this work was
done in 1965 and presents many images from instruments not
entirely operational, it is a valuable paper and highly rec-
ommended.

12. Belew, Leland F. SKYLAB, OUR FIRST SPACE STATION. NASA
SP-400. Washington, D.C.: U.S. Government Printing Office, 1977.
xi, 164 p. Figs., illus., index, col. photos.

Although Skylab, launched on 14 May 1973, performed some
earth remote-sensing observations, many of the mission objectives had little or nothing to do with the earth observation
program. This small booklet describes in great detail the
origin and execution of the Skylab program.

13. Bernstein, Ralph. ALL-DIGITAL PRECISION PROCESSING OF ERTS
IMAGES. Gaithersburg, Md.: International Business Machines Corp.,
April 1975. 149 p. Refs., illus., figs. Available from NTIS, N75-
20789.

Digital techniques have been developed and used to apply
precision-grade radiometric and geometric corrections to
ERTS MSS and RBV scenes. Geometric accuracies sufficient for mapping at 1:250,000 have been demonstrated.
Radiometric quality has been superior to ERTS NDPF precision
products normally supplied to a customer. A configuration
analysis has shown that feasible, cost-effective all-digital

systems for correcting ERTS data are easily obtainable. This report contains a summary of all the results obtained during this study (which was conducted for NASA, Goddard Space Flight Center under contract NAS5-21716) and includes: 1) radiometric and geometric correction techniques, 2) reseau detection, 3) GCP location, 4) resampling, 5) alternative configuration evaluations, and 6) error analysis. The report contains some excellent illustrations of ERTS data (color composite) as well as sufficient examples and procedures to be a very instructive aid for computer analysis of ERTS imagery.

14. Bodechtel, Johann, and Gierloff-Emden, Hans-Gunter. THE EARTH FROM SPACE. Translated from German by Hildegard Mayhew and Lotte Evans. New York: Arco; Newton Abott, Engl.: David and Charles, 1974. 176 p. Illus., glossary, figs., col. and b/w photos., maps.

The techniques of space surveys are discussed and a description is given of the special photographic methods employed on spacecraft. Over forty photos and fifty diagrams and maps (many in color) are included, thus making this publication one of the best presently available for relating the study of space photography to earth environments.

15. Bowker, David E., and Hughes, J. Kenrick. LUNAR ORBITER PHOTOGRAPHIC ATLAS OF THE MOON. NASA SP-206. Washington, D.C.: Scientific and Technical Information Office, U.S. National Aeronautics and Space Administration, 1971. v, 41 p. Figs., plates, tables, maps, photos. Available from U.S. Government Printing Office.

This atlas is an excellent selection of photography obtained from lunar orbiter spacecraft during 1966 and 1967. These photographs, from both the front and the back of the moon show greater surface detail than had previously been available from any source. Several maps of the moon, showing areas covered by each of the missions for low and high resolution, are included. Each of the 675 plates included in the atlas is accompanied by the photograph number, center coordinates, sun angle, and spacecraft altitude, together with a map showing the position of the area photographed. Scales of the photographs vary from 1 cm:5 km to 1 cm:195 km.

16. Branch, Melville C. CITY PLANNING AND AERIAL INFORMATION. Harvard City Planning Studies. Cambridge, Mass.: Harvard University Press, 1971. xvi, 283 p. Bibliog., col., and b/w photos.

This introductory textbook concentrates on photo interpretation, rather than photogrammetry, of urban scenes and situations as well as the application of aerial information to city planning tasks. A small section deals with other types of

General Literature

remotely sensed data (e.g., radar, spectrometer, interferometer, magnetometer). Some excellent examples of aerial
photography are included.

17. Bressanin, G., et al. DATA PREPROCESSING SYSTEMS FOR EARTH
RESOURCES. 5 vols. Paris: European Space Research Organization,
September 1973. Available from NTIS.

The aim of this study is to give earth scientists a panoramic
view of preprocessing methods and techniques for remotely
sensed data of the earth environment. The term "preprocessing" is used, rather than processing, to cover all of those
functions that convert observations into information. It is
realized that the boundary between preprocessing and processing is arbitrary; therefore, some of the techniques covered
in this review could very well be classified under the head-
ing of processing. Nonetheless, it was decided to retain
the term "preprocessing" thus not deviating from a termi-
nology already established in this field. This series consists
of the following five volumes.

Vol. 1.	INTRODUCTION TO PREPROCESSING TECHNIQUES	260 p.	N74–18055
Vol. 2.	METHODS OF IMPLEMENTATION	188 p.	N74–18056
Vol. 3.	TECHNICAL APPENDICES	91 p.	N74–18057
Vol. 4.	SUMMARY	31 p.	N74–18058
Vol. 5.	BIBLIOGRAPHY	134 p.	N74–18059

Also, the first volume provides basic information on remote
sensor operation and Vol. 5 (BIBLIOGRAPHY) contains an excellent and complete listing of works.

17A. Brosius, Charles A.; Gervin, Jannette C.; and Ragusa, James M. RE-
MOTE SENSING AND THE EARTH. Titusville, Fla.: School Board of
Brevard County, December, 1977. xxii, 460 p. Refs., figs., tables.
Available, NTIS N78-23509. Hard copy available from: School Board
of Brevard County, Instructional Services Division, Project Remote Sens-
ing, 1274 S. Florida Avenue, Rockledge, Fla. 32955.

This publication, possibly one of the best cooperative efforts
between NASA and a public school (high school and elementary levels) is oriented toward the study of remote-sensing
technology and techniques as they especially apply to the
ecological sciences. The publication is divided into four
major sections: Section One: "The Basics of Remote Sensing," in which the basic state-of-the-art of remote sensing
is presented; Section Two: "Selected Readings," consisting
of sixteen readings in the disciplines of agriculture, land
use, geology, water resources, marine resources, and the environment; Section Three: "Laboratory Excursions," a set of
fifteen laboratory exercises and, Section Four: "Appendices,"
of which there are eight. This is an excellent publication

and possibly the most comprehensive publication of its type available for and to the public school educators.

18. Colvocoresses, Alden P. REMOTE SENSING PLATFORMS. U.S. Geological Survey Circular 693. Washington, D.C.: U.S. Government Printing Office, 1974. vi, 75 p. Illus. Available from NTIS, N74-27825.

   Typical vehicles which carry remote sensors into the atmosphere or beyond are described and illustrated. Examples of airborne platforms and spacecraft are selected from those vehicles which have demonstrated acceptance and capability or have been defined for future remote-sensing missions. Except for a unique British kite balloon, only American platforms are covered. Remote sensing of the earth is the prime consideration, but sensing of other planets and moons is also considered.

19. Colwell, Robert N., et al. MONITORING EARTH RESOURCES FROM AIRCRAFT AND SPACECRAFT. NASA SP-275. Washington, D.C.: U.S. Government Printing Office, 1971. vii, 170 p. Illus., tables, refs. Available from NTIS, N72-18331.

   In view of the developing world population pressure and the needs for agricultural commodities, we need the wisest possible management of the earth's resources. An important step leading to such management is that of obtaining accurate resource inventories quickly and at frequent intervals. This report first describes an experiment that sought to determine the extent to which earth resources might be monitored by means of periodic inventories made with the aid of aerial and space photography. It then presents the results obtained from that experiment in each of several geographic test areas. Special emphasis is given to vegetation resources because (1) their intelligent management requires frequent monitoring and (2) the economic benefits are potentially very great. Consideration is given to the potential value of the experimental results to the resource manager and finally a chapter summarizes the experimental results, both quantitatively and qualitatively. Then, on the basis of these results, conclusions are drawn as to the advantages, limitations, and overall feasibility of monitoring earth resources with the aid of aerial and space photography. Eleven chapters, prepared by numerous people from diverse backgrounds, comprise the publication.

20. Cortright, Edger M., comp. and ed. EXPLORING SPACE WITH A CAMERA. NASA SP-168. Washington, D.C.: U.S. Government Printing Office, 1968. x, 214 p. Appendix, photos.

Representative space photographs from among the thousands of pictures returned to earth during the decade 1958–68 comprise this volume. The photographs are presented in sections devoted to photography above the atmosphere, to the moon and beyond, and man's venture into space. Some of the pictures were made by man-directed cameras, while others were relayed back to earth by unmanned spacecraft. Primarily a collection representing space accomplishments of the United States, the volume also includes some photographs taken during Soviet space probes. Cloud cover, storms, and other meteorological phenomena are shown in the photographs, for which written texts are included along with comments by NASA and other scientific personnel. Surveyor and Orbiter shots of the moon include photographs of craters, mares, and other formations. Shots from various Gemini flights depict the docking of the Agena target, man's space walk, earth's features, and space phenomena. The appendix includes details of the major spacecraft launched by the United States.

21. Darden, Lloyd. THE EARTH IN THE LOOKING GLASS. Garden City, N.Y.: Anchor Press, 1974. ix, 324 p. Illus., figs., index, bibliog., refs.

This book, written in a popular vein, presents numerous applications and uses of the several satellite systems which are presently surveying earth resources. The strongest emphasis is on the Landsat series. The author has included not only the uses of the data, but also the rationale and the needs for continued efforts.

22. Davis, R., et al., comps. USER DATA DISSEMINATION CONCEPTS FOR EARTH RESOURCES, FINAL REPORT. 2 vols. Prepared for NASA, Ames Research Center under contract NAS2-9864. Palo Alto, Calif.: Systems Control, June 1976. Figs., refs., tables. Available from NTIS, N76-33595, and N76-33596.

This study was directed toward the evaluation of domestic data dissemination networks for earth resources data in 1985–95. To do this the following topics were considered: data sources and volumes, future user demand, various types of transmission links, preprocessing requirements, network costs, and required technological developments to support this implementation. An interesting report which helps to illustrate some of the intricate planning that is involved in earth-resources satellite sensing systems is contained in these pages.

23. Earth Satellite Corp. EARTH RESOURCES SURVEY BENEFIT-COST STUDY: ECONOMIC, ENVIRONMENTAL, AND SOCIAL COSTS AND

BENEFITS OF FUTURE EARTH RESOURCES SURVEY SYSTEMS. 6 vols.
Prepared for the U.S. Geological Survey by Booz-Allen Applied Research. Washington, D.C.: November 1974. Var. pag. Available
from NTIS.

Volume 1 presents a summary of a study to evaluate the economic, environmental, and social costs and benefits of future
Landsat satellite systems. The results and conclusions of the
analysis, background of the study, and methods of analysis
are covered. The expected benefits in such applications
areas as agriculture, water resources, land use planning, and
rangeland management are summarized. Limitations of this
and other cost-benefit analyses as techniques for prediction
of real world results are also discussed. The remaining volumes and appendixes expand on these discussions. This report is most interesting and represents what is probably the
most complete analysis of the benefits and costs of the Landsat satellite system.

Contents:

Vol. 1.	EXECUTIVE SUMMARY	51 p.	N75-20813
Vol. 2.	SUMMARY OF BENEFITS EVALUATIONS	323 p.	N75-20814
Vol. 3.	ALTERNATE SYSTEMS EFFECTIVENESS ANALYSIS	175 p.	N75-20815
Vol. 4.	CAPABILITIES TO DERIVE INFORMA- TION OF VALUE FROM ERS DATA	346 p.	N75-20816
Vol. 5.	APPROACH AND METHODS OF ANALYSIS	102 p.	N75-20817
Vol. 6.	ANALYSIS OF DISTRIBUTIONAL, ENVIRONMENTAL, SOCIAL, AND INTERNATIONAL IMPACTS	172 p.	N75-20818
Appendix 1.	An Analysis of the Benefits and Costs of an Improved Crop Acreage Forecasting System Utilizing Earth Resources Satellite or Aircraft Information	147 p.	N75-20819
Appendix 2.	Snow Mapping and Runoff Fore- casting: Examination of ERTS-1 Capabilities and Potential Benefits from an Operational ERS System	229 p.	N75-20820
Appendix 3.	Rangeland Case Study	242 p.	N75-20821
Appendix 4.	An Analysis of the Benefits and Costs in Forestry Utilizing Earth Resources Satellite or Aircraft Information	213 p.	N75-20822
Appendix 5.	An Analysis of Costs and Benefits from Use of ERS Data in State Land Use Planning	298 p.	N75-20823
Appendix 6.	An Analysis of the Benefits and Costs from the Use of ERS Data in Environmental Analysis	145 p.	N75-20824

Appendix 7. Living Marine Resources Broad Area   28 p.   N75-20825
             Analysis

24.   Eastman Kodak Co. APPLIED INFRARED PHOTOGRAPHY. Kodak Pub-
       lication M-28. Rochester, N.Y.: 1972. 88 p. Charts, figs.,
       plates. Paper.

       Following a discussion of the basic understanding of IR pho-
       tography, is an important section about the use of IR out-
       doors. Numerous examples and a clear text make this valu-
       able reading for earth scientists interested in remote sensing.
       Equipment and materials are described as are numerous labo-
       ratory techniques and applications of IR photography in areas
       other than earth-resources remote sensing.

25.   ECON, Inc. SEASAT ECONOMIC ASSESSMENT. 10 vols. Prince-
       ton, N.J.: August 1975. Available from NTIS.

       This report covers the period of February 1974 to August
       1975 and consists of ten volumes which deal with economic
       assessment, utility of SEASAT data, trade-off possibilities,
       system description and performance characteristics and case
       studies. The study was conducted for NASA. These pre-
       launch studies (SEASAT was launched in mid-1978) are
       especially interesting in that they help to show a portion of
       the process through which such a complex system must be
       studied to ensure a successful launch.

       The volumes are listed below:

       Vol. 1. SUMMARY AND CONCLUSIONS   34 p.   N76-28614
       Vol. 2. SEASAT DESCRIPTION AND   99 p.   N76-28615
               PERFORMANCE
       Vol. 3. OFFSHORE OIL AND NATURAL GAS   147 p.   N76-28616
               INDUSTRY CASE STUDY AND
               GENERALIZATIONS
       Vol. 4. OCEAN MINING CASE STUDY AND   43 p.   N76-28617
               GENERALIZATION
       Vol. 5. COASTAL ZONES CASE STUDY AND   91 p.   N76-28618
               GENERALIZATION
       Vol. 6. ARCTIC OPERATIONS CASE STUDY   77 p.   N76-28619
               AND GENERALIZATION
       Vol. 7. MARINE TRANSPORTATION CASE   289 p.   N76-28620
               STUDY
       Vol. 8. OCEAN FISHING CASE STUDY   131 p.   N76-28621
       Vol. 9. PORTS AND HARBORS CASE STUDY   201 p.   N76-28622
               AND GENERALIZATION
       Vol. 10. THE SATIL 2 PROGRAM (A Program 206 p.   N76-28623
               for the Evaluation of the Costs of an
               Operational SEASAT System as a Function
               of Operational Requirements and Reliability)

26. Elliot Automation Space and Advanced Military Systems. SIDE-LOOK-ING RADAR SYSTEMS AND THEIR POTENTIAL APPLICATIONS TO EARTH RESOURCES SURVEYS. 7 vol. Camberley, Engl.: 1972-73. Figs., appendix, refs.

This report is one of a set prepared under contract from ESRO with the purpose of informing the earth scientists on the applications of side-looking radar to the remote sensing of earth resources. The set also indicates the required development, support program, and costs necessary for the European capability in remote sensing using SLR. The complete lists of reports is as follows:

Vol. 1.	BASIC PRINCIPLES OF SIDE-LOOKING RADAR	R.A. Dean, ed.	August 1972 vii, 86 p. N73-12418
Vol. 2.	RADAR SCATTERING FROM NATURAL SURFACES	R.A. Dean and A.R. Domville, eds.	August 1972 v, 25 p. N73-12153
Vol. 3.	POTENTIAL APPLI-CATIONS OF SLR TO THE REMOTE SENSING OF EARTH RESOURCES	R.A.G. Savigear et al	July 1972 vl, 135 p. N73-12401
Vol. 4.	DATA PROCESSING AND INTERPRETA-TION REQUIREMENTS	R.A. Dean and R.J. Haslam	August 1972 vii, 100 p. N73-12419
Vol. 5.	RECOMMENDATIONS FOR A EUROPEAN PROGRAMME	Unknown	
Vol. 6.	SUMMARY	K. Grant et al.	March 1973 xvi, 136 p. N73-15200
Vol. 7.	BIBLIOGRAPHY	R.A. Dean	August 1972 iii, 100 p. N73-15198

27. Estes, John E., and Senger, Leslie W., eds. REMOTE SENSING: TECHNIQUES FOR ENVIRONMENTAL ANALYSIS. Santa Barbara, Calif.: Hamilton Publishing Co., 1974. 340 p. Illus., tables, figs., glossary, refs.

This publication is designed as a reference book and is an outgrowth of reference readings for a remote-sensing workshop in conjunction with the twenty-fourth International Geograph-ical Union meeting in Montreal, Canada, in August 1972. It contains a good index and listing of institutions and organi-zations engaged in remote-sensing research and applications. Qualitative data extraction and the analysis of remote sensor images, imaging with photographic and nonphotographic sensor systems, and geomorphic-geologic mapping from data obtained by remote sensors are among the topics covered in the papers presented. Other topics covered include interpretation and mapping of natural vegetation, remote sensing of agricultural resources, and urban applications of remote sensing.

28. Eynard, Raymond A., ed. COLOR: THEORY AND IMAGING SYSTEMS. Washington, D.C.: Society of Photographic Scientists and Engineers, 1973. 448 p.

The theory and history of color photography are described in sections dealing with classical processes, significant breakthroughs, inventions, and modern imaging systems. Attention is given to the response of the human eye, human color perception, colorimetry, color densitometry, color sensitometry, masking, color reproduction in color television, chemical fundamentals of color development, color reversal silver halide systems, silver dye bleach, color copy materials, color films for aerial photography, generation of color imagery from ERTS data, color microfilming, and color zerography.

29. Fischer, William A.; Hemphill, W.R.; and Kover, A. "Progress in Remote Sensing (1972-1976)." PHOTOGRAMMETRIA 32 (1976): 33-72.

This paper is possibly one of the most current reviews of the state-of-the-art of the applications of remote sensing. The authors note that, during the four years covered, major research thrusts have been: (1) computer assisted enhancement and interpretation, (2) earth science applications for Landsat data, and (3) investigations of the usefulness of observations of luminescence, thermal infrared, and microwave energies. They also note that thermal surveys from aircraft have become largely operational, but continued research in thermal modeling and analysis of high altitude data is occurring. Microwave research is increasing rapidly and programs are being developed for satellite observations. Microwave research concentrating on oil spill detection, soil moisture measurement, and observations of ice distributions have also received considerable attention.

30. General Electric Co. Space Division. Advanced NASA Programs Division. DEFINITION OF THE TOTAL EARTH RESOURCES SYSTEM FOR THE SHUTTLE ERA (TERSSE). 11 vols. Conducted under NASA Contract NAS 9-13901. Valley Forge, Pa.: 1974-75. Available from NTIS.

The pressing need to survey and manage the earth's resources and environment, to better understand remotely sensed phenomena, to continue technological development, and to improve management systems are all elements of a future earth resources system. The Space Shuttle brings a new capability to earth resources survey including direct observation by experienced earth scientists, quick reaction capability, spaceborne facilities for experimentation and sensor evaluation, and more effective means for launching and servicing long mission life space systems. The Space Shuttle is, however, only one element in a complex system of data gathering, translation, distribution, and utilization functions. While the Shuttle more decidedly has a role in the total earth resources program, the central question is the form of the future earth resources system itself. It is only by analyzing this form and accounting for all elements of the system that the proper role of the Shuttle in it can be made visible. This study, entitled TERSSE, Total Earth Resources System for the Shuttle Era, was established to investigate the form of this future earth resources system. The constituent system elements of the future earth resources system and the key issues which concern the future earth resources program are both complex and interrelated in nature. The purpose of this study has been to investigate these items in the context of the total system utilizing a comprehensive, systems-oriented methodology. This series of reports consists of the following:

Executive Summary		March 1975	29 p.	N75-31552
Vol. 1.	EARTH RESOURCES PROGRAM SCOPE AND INFORMATION NEEDS	November 1974	443 p.	N75-31544
Vol. 2.	AN ASSESSMENT OF THE CURRENT STATE-OF-THE-ART	October 1974	239 p.	N75-31545
Vol. 3.	MISSION AND SYSTEM REQUIREMENTS FOR THE TOTAL EARTH RESOURCES SYSTEM	November 1974	258 p.	N75-31546
Vol. 4.	THE ROLE OF THE SHUTTLE IN THE EARTH RESOURCES PROGRAM	November 1974	134 p.	N75-31547
Vol. 5.	DETAILED SYSTEM REQUIREMENTS: TWO CASE STUDIES	November 1974	310 p.	N75-31548
Vol. 6.	AN EARTH SHUTTLE PALLET CONCEPT FOR THE EARTH RESOURCES PROGRAM	November 1974	70 p.	N75-31549
Vol. 7.	USER MODELS: A SYSTEM ASSESSMENT	October 1974	91 p.	N75-31550
Vol. 8.	USER'S MISSION AND SYSTEM	October 1974	303 p.	N75-31551

REQUIREMENT DATA.
(APPENDIX A OF
VOLUME 3)

Vol. 9. EARTH RESOURCES   August 1975   170 p.   N76-29686
SHUTTLE APPLICA-
TIONS

Vol. 10. TOSS--TERSSE OPER- December 1975   626 p.   N76-29687
ATIONAL SYSTEMS
STUDY

31.   Gonin, G.B., et al., eds. SPACE PHOTOGRAPHY AND GEOLOG-
ICAL STUDIES   [Kosmicheskaya fotos'yemka i geologicheskiye issledo-
vaniya]. Translated from the Russian by L. Kanner Associates, Redwood
City, Calif. NASA Technical Trans. F-16852. Washington, D.C.:
National Aeronautics and Space Administration, February 1976. 521 p.
Illus., refs. Available from NTIS, N76-19529.

> This book deals with theoretical and practical problems of
> space photography and geologic interpretations of space pho-
> tography. Obtaining images of the sunlit surface of the
> earth from spacecraft is examined, involving spacecraft ori-
> entation and laws of displacement deriving from celestial
> mechanics. Stress is placed on the effect of the atmosphere
> on the quality of the imagery. Results of satellite-track ex-
> periments involving the simultaneous photographing of several
> areas of the earth from space and from aircraft are presented,
> along with data on aircraft measurements of optical charac-
> teristics of certain landscape objects through significant at-
> mospheric thickness. Space photogrammetry is represented by
> a technique for the rigorous solution of inverse photogram-
> metric intersection of space photographs, analysis of their
> measuring properties, transmission of space photographs, and
> compilation of photomaps in a given projection from space
> photographs. Also treated is stereophotogrammetry of space
> photographs. The publication contains 537 references.

32.   Goodyear Aerospace. DEVELOPING EARTH RESOURCES WITH SYN-
THETIC APERATURE RADAR. Code GIB-9290E. Litchfield Park, Ariz.:
1975. 35 p. Illus., figs., maps. Paper, (spiralbound).

> This booklet presents an excellent series of annotated glossy
> reproductions of Side Looking Airborne Radar (SLAR) imagery.
> Applications specifically addressed are as follows: map com-
> pilation and updating, ice surveillance, geology, hydrology,
> forestry and agriculture, land use planning, and natural dis-
> aster assessment. Scales for the presented imagery vary from
> 1:50,000 to 1:750,000. This is the most recent in a series
> of similar booklets, listed below:

NONMILITARY APPLICATIONS Code GIB-9290B   35 p.   March 1973
OF SYNTHETIC APERTURE
     RADAR

EARTH RESOURCES APPLICA-   Code GIB-9290C   35 p.   March 1973
TIONS OF SYNTHETIC
APERTURE RADAR

DEVELOPING EARTH RE-   Code GIB-9290D   35 p.   August 1973
SOURCES WITH SYNTHETIC
APERTURE RADAR

33. Grady, James. PHOTO-ATLAS OF THE UNITED STATES. Pasadena,
Calif.: Ward Ritchie Press, 1975. 127 p. Photos.

> Two sets of photomaps based on Landsat and Skylab or U-2
> imagery are included in this publication. The former are
> printed primarily in 'earth blue' at the scale of 1:140,500,
> the latter in color generally at unknown scales. Many of
> the photographs have major geographic and topographic fea-
> tures named.

34. Grigor'yev, A.A. SPACE REMOTE SENSING OF THE EARTH LAND-
SCAPES [Kosmicheskaya indikatsiya landshaftov zemli]. Translated
from the Russian by Agnew Tech-Trans, Woodland Hills, Calif. NASA
Technical Trans. F-16924. Washington, D.C.: U.S. National Aero-
nautics and Space Administration, 1975. 169 p. Illus., index, refs.
Available from NTIS, N76-27644.

> The development of space photography has opened new and
> wide possibilities for the investigation of landscapes which
> are important for solving meteorological problems and for the
> development of space geography. This book, based both on
> original investigations of the author and the experience of
> Soviet and foreign researchers, considers in a systematized
> manner the possibilities of the remote sensing of various types
> of the earth's underlying surfaces. Various types of space
> imageries from different regions of the world are analyzed.
> This book is designed for people studying questions of remote
> sensing of the environment and may be used by space ecology
> specialists, meteorologists, geophysicists, geographers, and so
> forth. A total of 362 references and a subject index are
> included.

35. Haralick, Robert M. "Glossary and Index to Remotely Sensed Image
Pattern Recognition Concepts." PATTERN RECOGNITION 5, no. 4
(December 1973): 391-403.

> This glossary, the purpose of which is to state in the simplest
> possible way the general meaning or word usage for many of
> the terms in image pattern recognition, is oriented toward
> those readers who are generally unfamiliar with the area and
> to provide them with an overall perspective of the field.
> Over one hundred terms are included in the glossary which
> was prepared as the report from the definitions and standards

subcommittee, the Automatic Image Pattern Recognition Committee of the Electrical Industries Association. Mathematical
formulae involving integrals or derivatives are not included
in the definitions.

36. Harger, Robert O. SYNTHETIC APERATURE RADAR SYSTEMS: THEORY
    AND DESIGN. Electrical Sciences: A Series of Monographs and
    Texts. New York: Academic Press, 1970. xiii, 240 p. Illus.,
    figs., tables, refs.

    This is the first book that is devoted to the principles, the-
    ory, and system design of synthetic aperture radar (SAR) sys-
    tems. Although it is rather detailed and designed for the
    student in electrical engineering, it can provide needed
    background and detailed statements about the SAR systems
    which are helpful for people using radar remote sensing for
    earth-resources studies.

37. Harper, Dorothy. EYE IN THE SKY: INTRODUCTION TO REMOTE
    SENSING. Montreal: Multiscience Publications, 1976. 172 p.
    Figs., illus., tables. Paper.

    This short book, oriented toward the senior high school and
    junior college student, is an excellent statement about EM
    remote sensing, its nature, and its applications. Also,
    Harper's book provides a background for understanding Ca-
    nadian contributions and involvement in remote sensing.

38. Hazelrigg, George A., Jr., and Heiss, Klaus P., project directors.
    THE ECONOMIC VALUE OF REMOTE SENSING OF EARTH RESOURCES
    FROM SPACE: AN ERTS OVERVIEW AND THE VALUE OF CONTIN-
    UITY OF SERVICE. 10 vols. Princeton, N.J.: ECON, Inc. De-
    cember 1974. Available from NTIS.

    The Earth Resources Technology Satellite (ERTS) program faces
    some crucial decisions over the next twelve to eighteen
    months that will affect the future of remote sensing by sat-
    ellites for decades to come. The purpose of this report is
    to provide an overview of the ERTS program to date, to de-
    termine the magnitude of the benefits that can be reasonably
    expected to flow from an Earth Resources Survey (ERS) pro-
    gram, and to assess the benefits foregone in the event of a
    one-year or a two-year gap in ERS services occurring in
    1977-78. The ten volumes (NTIS nos. N75-14203 to N75-
    14215 inclusive) cover the following areas: agriculture, for-
    estry, inland water resources, land use, nonreplenishable nat-
    ural resources, atmosphere, oceans, and industry.

39. Heaslip, George B. ENVIRONMENTAL DATA HANDLING. A Wiley-
    Interscience Publication, Environmental Science and Technology Series.

New York: J. Wiley and Sons, 1975. 203 p. Illus., figs., tables, glossary, refs.

Designed as a text for the fundamentals of remote-sensor data acquisition and data handling, this publication provides a background which is oriented in the engineering and physics realm. A very useful book for conceptual understandings of the basics of sensor operation and some data analysis options. Few examples of applications are given.

39A Heiman, Grover. AERIAL PHOTOGRAPHY: THE STORY OF AERIAL MAPPING AND RECONNAISSANCE. Air Force Academy Series. New York: Macmillan Co., 1972. xii, 180 p. Illus., index.

As the name implies, this book is primarily a history of the reconnaissance program of the U.S. Air Force. Written in a style which is easily read and understood this book can provide an interesting background and "broadening" for those interested primarily in reconnaissance aerial photography. Some small portion is allocated to nonphotographic sensors.

40. Hewish, Anthony, ed. SEEING BEYOND THE VISIBLE. New York: American Elsevier Publishing Co.; London: English Universities Press, 1970. vii, 150 p. Figs., plates.

This book contains a series of seven lectures originally presented as a BBC series titled AT THE SPEED OF LIGHT. Written for the layman from the point of view of physicists and astronomers, it presents an excellent understanding of the EM system and a good background for persons working in the area of earth-resources remote sensing.

41. Holter, Marvin R., et al. FUNDAMENTALS OF INFRARED TECH-NOLOGY. Macmillan Monographs in Applied Optics. S.S. Ballard, consulting editor. New York: Macmillan Co., 1962. iv, 442 p. Tables, figs., index.

This book is an outgrowth of the notes used by the authors in a one-week accelerated course in modern infrared technology. Thus, although oriented primarily toward engineers, equipment designers, and physicists, it provides a very substantial background for the earth scientists wishing to consider applications of these technologies.

42. Holz, Robert K., ed. THE SURVEILLANT SCIENCE: REMOTE SEN-SING OF THE ENVIRONMENT. Boston: Houghton Mifflin Co., 1973. 390 p. Illus., bibliog., refs. Paper.

This book is a series of forty-four articles covering the field of remote sensing and chosen because they are good examples of research and writing in the specific areas of remote sensing.

Each article is supplied with an editor's introduction which provides transition between the articles.

43. Horton, Frank E. THE APPLICATION OF REMOTE SENSING TECHNIQUES TO INTER AND INTRA URBAN ANALYSIS. U.S. Geological Survey Interagency Report 250. Iowa City: University of Iowa, May 1972. 254 p. Paper. Available from NTIS, PB-214-449.

This is an effort to assess the applicability of air and spaceborne photography toward providing data inputs to urban and regional planning, management, and research. Through evaluation of remote-sensing inputs to urban change detection systems, it is determined that remote sensing can provide data concerning: (1) land use, (2) changes in commercial structure, (3) data for transportation planning, (4) housing quality, (5) residential dynamics, and (6) population density.

44. International Union of Forest Research Organizations. APPLICATIONS OF REMOTE SENSORS IN FORESTRY. A joint report with the Working Group on Applications of Remote Sensors in Forestry. Frieburg, Germany: Druckhaus Rombach and Co., 1971. 189 p. Figs., tables, refs.

This book, consisting of thirteen papers, is an expansion of the papers originally presented in Section Twenty-five of the Plenary Session report of the XV IUFRO Congress held in 1971. Many remote-sensing systems are discussed, but generally the authors of the several papers are of the opinion that aerial photography will continue to be one of the major sensors in these areas of study.

45. Karr, Clarence, Jr., ed. INFRARED AND RAMAN SPECTROSCOPY OF LUNAR AND TERRESTRIAL MINERALS. New York: Academic Press, 1975. xii, 375 p. Figs., illus., index.

"The purpose of this volume is to make available in a single reference work original descriptions and summaries of the research on infrared and Raman spectroscopy of lunar and terrestrial minerals so that this information will be readily available to researchers . . . in particular to that group of researchers in government, industry and universities involved in the many programs on terrestrial minerals and earth sciences by remote sensing." The chapters in the book are arranged according to spectroscopic technique and/or frequency range, with chapters 4 through 7 dealing primarily with remote sensing in the visible and infrared, both reflectance and emittance of terrain.

46. Kennedy, J.M., et al. PASSIVE MICROWAVE MEASUREMENTS OF SNOW AND SOIL. 2 vols. El Monte, Calif.: Advanced Microwave Systems Division, Space-General Corp., November 1966. Illus., figs., refs. Available from NTIS, N67-24793, and N67-24785.

> The basic objectives of the program were to measure the radiometric brightness temperatures of snow, sand, soil, mud, and rock both in the field and in the laboratory. This was later expanded to include beach sands, marine swamp, and playa lake beds. These measurements were then related to the physical parameters, changes in the environment, and sensor frequencies. Data obtained on in situ material confirmed that moisture content is an important parameter, probably the most important one, affecting radiometric brightness temperatures. The effect of water seems to be independent of the base material, so that increasing the free water content has a similar effect whether the base material is snow, soil, or sand. These two volumes are of importance because they were some of the early works detailing the microwave brightness temperatures of various earth surface features in a detailed and systematic way, and because an understanding of these temperatures is necessary for a proper interpretation of Nimbus satellite series data.

47. _____. PASSIVE MICROWAVE MEASUREMENTS OF SNOW AND SOIL: A STUDY OF THE THEORY AND MEASUREMENTS OF THE MICROWAVE EMISSION PROPERTIES OF NATURAL MATERIALS. Done for the Geography Branch, Earth Science Division, Office of Naval Research. El Monte, Calif.: Advanced Microwave Systems Division, Space-General Corp., February 1967. 64 p. Figs., illus., refs. Available from NTIS, N67-26907, and AD 648 818.

> The first part of the report is a general review of techniques for measuring dielectric constant of materials at frequencies in the millimeter and centimeter regions of the EM spectrum. Following a review of various systems for measuring dielectric constants, the relationship between dielectric constants and power reflection coefficients is developed. In the final section, the relationships of the dielectric constant to the physical conditions of snow is fully developed. There is a unique dielectric constant for snow when the melt condition and density are specified. Using this relation and the relationship between dielectric constant and Fresnel reflection coefficients, radiometric temperature, polarization, and angle observation may be used to fully analyze snow from a remote platform.

48. Kock, Winston E. RADAR, SONAR, AND HOLOGRAPHY: AN INTRODUCTION. New York: Academic Press, 1973. xvii, 140 p.

> "The purpose of this little book is to provide an introduction

to the technology of radar and sonar. Because the new science of holography . . . is affecting both of these fields
quite strongly, the book includes an explanation of the fundamental principles underlying this new art . . . and of the
hologram process itself. Finally, numerous examples are discussed which show how holography is providing new horizons
to radar and sonar systems. The book thus also provides a
simple approach to the new technology of holography. It is
therefore hoped that the discussions to follow will make clear
the basic difference which exists between photography and
holography . . . and between standard sonar and radar."
This is a clearly written and easily understood book which
provides much basic background information for understanding
earth-resources imaging radars.

49. Kong, J.A., ed. THEORY OF PASSIVE REMOTE SENSING WITH
MICROWAVES: FINAL REPORT. Cambridge: Massachusetts Institute
of Technology, Research Laboratory of Electronics, 15 July 1975.
301 p. Figs., tables, refs. Available from NTIS, N76-18629.

This publication consists of two major sections. The first,
consisting of twenty-seven pages, presents the general theory
of remote sensing with microwaves. The second section contains six appendixes consisting of a series of reports (some
presented for bachelor of science degrees at M.I.T.) dealing
with specific aspects of passive microwave remote sensing.

50. Kosofsky, L.J., and El-Baz, Farouk. THE MOON AS VIEWED BY
LUNAR ORBITER. NASA SP-200. Washington, D.C.: Office of
Technology Utilization, National Aeronautics and Space Administration,
1969. vii, 152 p. Plates, figs., appendixes, maps. Spiralbound.
Available from NTIS.

This excellent and selected compilation of lunar photography
from spacecraft consists of three major chapters: (1) "Introduction," (2) "A Distant View," and (3) "A Close Up View."
Within chapter 3, the features are identified in four major
groups of maria, highlands, craters and faults, rills and
domes. The photography is selected from five lunar missions.
The appendixes consist of four-color stereoscopic views, eight
index maps, and photo references.

51. Kovaly, John J. SYNTHETIC APERTURE RADAR. Dedham, Mass.:
Artech House, 1976. x, 333 p. Illus., figs., refs.

This is one of the most recent books in a field of remote
sensing. Consisting of seven chapters and thirty-three reprinted articles (primarily from the IEEE publications) this
book is oriented toward engineering rather than the applications of SAR. The chapters included are as follows: (1)
"Development of SAR," (2) "Comparative SAR Theory,"

(3) "Performance of SAR," (4) "Effect of Random Phase Errors on Synthetic Arrays," (5) "Motion Compensation in SAR," (6) "Processing for SAR" and (7) "Applications of SAR."

52. Kudritskii, D.M.; Popov, I.V.; and Romanov, E.A. HYDROGRAPHIC INTERPRETATION OF AERIAL PHOTOGRAPHS [Osnoy gidragraficheskogo deshifriovaniya aerofotosnimkov.] Translated by Israel Program for Scientific Translations. Washington, D.C.: U.S. Department of Commerce and National Science Foundation, 1966. x, 279 p. Refs., figs. Available from NTIS, TT64-11094.

> The State Hydrological Institute conducted a number of studies for the purpose of developing the use of the materials of aerial photography for descriptive hydrographic work, for information on the descent of snow cover and for the clarification of some special problems. As a result of this work it was decided that the wide use of the available material of aerial photography for practical hydrographic work should be made. Hence, the present book was prepared and should serve as a practical handbook for engineer hydrologists who use aerial photography in hydrological work. The book is divided into three major sections: (1) general information on aerial photography and methods of interpreting aerial photographs, (2) hydrographic interpretation of aerial photographs, and (3) hydrographic interpretation of marshes on aerial photographs. This is possibly one of the most complete single publications on this subject.

53. Lee, Willis Thomas. THE FACE OF THE EARTH AS SEEN FROM THE AIR: A STUDY IN THE APPLICATION OF AIRPLANE PHOTOGRAPHY TO GEOGRAPHY. American Geographical Society Special Publication no. 4, edited by W.L.G. Joerg. New York: American Geographical Society, 1922. ixx, 110 p. Illus., index.

> This book is most interesting because it is one of the earliest references generally available which deals with the application of aerial photography to earth resources. Although the discipline has made many advances in the past fifty-seven years, it is interesting to note that many of the comments made nearly six decades ago are still pertinent to the study of aerial photography. Photographs are both oblique and vertical and were gleaned primarily from the files of the U.S. Army Air Service, the U.S. Geological Survey, and the U.S. Navy Air Service. A wide range of earth features, both cultural and physical, are included in the examples given.

54. Levine, Daniel. RADARGRAMMETRY. New York: McGraw-Hill Book Co., 1960. xxi, 330 p. Illus., figs., charts.

With the increased importance of radar as a remote sensor (e.g., SEASAT, Space Shuttle), an increasing number of efforts will be made to use radar as a source of information for cartographic work and other precise measurements of the earth's surface. This book, although dealing primarily with airborne PPI radar systems, presents a theory and numerous examples of obtaining measurements from radar data, primarily imagery. Also included are very excellent discussions of the operation of the radar set--necessary discussions to help the interpreter understand some of the variations observed on the imagery which are system, rather than earth-resources, related.

55. Lintz, Joseph, Jr., and Simonett, David S., eds. REMOTE SENSING OF ENVIRONMENT. Reading, Mass.: Addison-Wesley Publishing Co., 1976. xx, 694 p. Illus., figs., tables, refs.

Remote sensing is the detection from a distance of physical features upon or near the earth's surface. By utilizing aircraft and spacecraft as platforms, new types of information can be provided for increased efficiency in the development and management of the earth's natural resources. This publication, a multidisciplinary, essentially nonmathematical textbook, is designed for first-year graduate students and advanced undergraduate students interested in natural resource areas such as agriculture, forestry, geography, geology, hydrology, and oceanography. The volume comprises the work of twenty contributors, and is divided into four parts which concern fundamental principles, sensors for remote sensing, processes and techniques, and applications to the natural resources. It offers illustrations in color and over nine hundred references.

56. Lo, C.P. GEOGRAPHICAL APPLICATIONS OF AERIAL PHOTOGRAPHY. New York: Crane, Russak, 1976. 330 p. Figs., plates, bibliog., index.

This publication, intended for advanced geographers who have some background in photogrammetry but lack practical experience, presents a statement of the state-of-the-art of aerial photography. Numerous references and comprehensive illustrations, together with many examples of applications of aerial photography and the techniques of the field make this a very useful book for reference and research needs.

57. Long, Maurice W. RADAR REFLECTIVITY OF LAND AND SEA. Lexington, Mass.: D.C. Heath and Co., Lexington Books, 1975. xix, 366 p. Refs., figs., illus., index.

This publication has brought together under one cover the engineering studies concerning reflectivity of radar signals (backscatter) and terrain classification information. Although somewhat preliminary from the point of view of many earth scientists, it provides a valuable starting point for bridging the often seen gap between engineering studies and earth applications.

58. Lowe, D.S.; Summers, R.A.; and Greenblat, E.J. AN ECONOMIC EVALUATION OF THE UTILITY OF ERTS DATA FOR DEVELOPING COUNTRIES. 2 vols. Report 105100-8-F. Ann Arbor: Environmental Research Institute of Michigan, August 1974. Available from NTIS, N75-16404, and N75-16405.

The utilization of ERTS-1 data in eighteen developing countries is reviewed and evaluated. This overall assessment is supported by more detailed economic evaluations in selected countries. Two quantitative economic evaluations were conducted, one on the benefits to be derived from improved rice-crop forecasting in Thailand and the other on benefits stemming from improved range-carrying capacity estimations in Kenya. Additional qualitative evaluations were made of potential improvements in mineral exploration, water resources management, and cartographic mapping. These benefits can be accrued at acceptably low cost levels, provided that appropriately timed and scaled technical assistance is provided. Volume 1 is titled FINAL REPORT, volume 2 consists of the appendixes.

59. Lowman, Paul D., Jr. SPACE PANORAMA. Feldman/Zurich, Switzerland: Weltflugbild, Reinhold A. Müller, 1968. 164 p. Illus.

This book is a collection of some of the best terrain photography obtained by the Gemini experiments. A total of sixty-eight photographs, together with annotations and discussions are presented in six sections (North America, South America, Africa, Asia, Australia, and the oceans). NASA photo identification numbers are given for all photographs, consequently, it is possible to easily order these using the proper catalog listed elsewhere in this bibliography.

60. Lueder, Donald R. AERIAL PHOTOGRAPHIC INTERPRETATION: PRINCIPLES AND APPLICATIONS. McGraw-Hill Civil Engineering Series. H.E. Davis, consulting editor. New York: McGraw-Hill Book Co., 1969. xv, 462 p. Illus., bibliog.

This book is an excellent text dealing with the elements of aerial photograph interpretation especially as applied to the earth sciences. It discusses photogrammetry as well as interpretation, the collection of aerial data, and instrumentation.

61. Marshall, Ernest W. AIR PHOTO INTERPRETATION OF GREAT LAKES

ICE FEATURES. Great Lakes Research Division. Special Report no. 25. Ann Arbor: Institute of Science and Technology, University of Michigan, 1966. ix, 92 p. Illus., refs., glossary.

> The report contains visual imagery of ice features and patterns common to the Great Lakes and gives interpretations. Ice features are included from all the Great Lakes with the majority from Lake Erie. The photographs and interpretations are arranged according to features found in open water areas during freeze-up, in the newly formed ice, in winter ice, and in patterns resulting from snow and wind. The report brings out the role of snow and water turbulence in determining the types of ice sheet formed. Recommendations are made for future research.

62. Michigan, University of. Institute of Science and Technology. Willow Run Laboratories. PEACEFUL USES OF EARTH OBSERVATION SPACE-CRAFT. Prepared for NASA under contract NASw-1084. 3 vols. Washington, D.C.: National Aeronautics and Space Administration, September 1966. Refs. Available from NTIS, N66-37028 to N66-37030 inclusive.

> A major objective of the program of NASA is to investigate and implement the adaptation of space technology for peaceful uses. As part of this program a comprehensive study of the requirements for conducting an integrated set of experiments in a series of manned earth-orbiting laboratories which would lead to the realization of such peaceful uses of space has been conducted. In addition a three-month study to survey potential applications of observation spacecraft in a number of important scientific and economic activities and to consider the program of ground-based and orbital experiments required to develop this capability has been conducted. These studies have resulted in this report. The results of the first phase of this investigation are reported in this three volume series. Volume 1 is an INTRODUCTION AND SUMMARY of the work performed, volume 2, SURVEY OF APPLICATIONS AND BENEFITS, contains a comprehensive survey of potential applications of earth-observation spacecraft and the anticipated benefits, and volume 3, SENSOR REQUIREMENTS AND EXPERIMENTS, describes some of the requirements to be met by the orbiting sensing devices and the manned earth-orbiting experiments proposed for developing the orbital sensing capability. Although this report is now of limited immediate value, it does provide a historical perspective of the origins of some of the present earth orbital remote-sensing programs.

63. Mutch, Thomas A., et al. THE GEOLOGY OF MARS. Princeton, N.J.: Princeton University Press, 1976. ix, 400 p. Figs., refs., illus.

Although this reference guide is oriented toward earth resources remote sensing, this publication on the geology of Mars is included because it is quite interesting in terms of remote sensing in general. Because we have no samples of the surface of Mars, it is necessary that all information about this planet be obtained remotely and by using analogs from the earth. Thus, many approaches from the most elementary observations to the most erudite pronouncements are found in this book. It is an excellent example of what can be inferred and concluded from remotely sensed data--although admittedly, we have not been able to empirically confirm the results of the research.

64. National Academy of Sciences, Washington, D.C. Committee on Remote Sensing Programs for Earth Resources Surveys. REMOTE SENSING FOR RESOURCE AND ENVIRONMENTAL SURVEYS: A PROGRESS REVIEW--1974. Washington, D.C.: Commission on Natural Resources, National Research Council, National Academy of Sciences, 1974. 109 p. Available from NTIS, N75-18705.

A committee report on resources and environmental information extracted from ERTS data is discussed along with problems faced by users of such information. Special attention is given to the following problems: (1) lack of assurance that the program will be continued beyond the technology demonstration phase and (2) the strength of repetitive synoptic space imagery, with selective spectral range and resolution but with lower spatial resolution, does not readily fit into the information process and decision models currently used by many operational managers.

65. National Academy of Sciences, Washington, D.C. PRACTICAL APPLICATIONS OF SPACE SYSTEMS. 13 vols. Washington, D.C.: National Academy of Sciences, National Research Council, 1975. Available from NTIS.

This is a series of reports (individual papers are referred to as supporting papers) which presents the findings and recommendations of assorted panels for developing a satellite remote-sensing global information system in the next decade. User requirements are often identified and potential benefits are discussed in most of the volumes. All volumes are available on microfiche from NTIS as noted below:

Panel				
1.	WEATHER AND CLIMATE	35 p.	N76-18749
2.	USE OF COMMUNICATIONS	55p.	N76-18301
3.	LAND USE PLANNING	65 p.	N76-18623
4.	AGRICULTURE, FOREST AND RANGE	57 p.	N76-18624
5.	INLAND WATER RESOURCES	87 p.	N76-18625
6.	EXTRACTABLE RESOURCES	32 p.	N76-18626

7.	ENVIRONMENTAL QUALITY	66 p.	N76-18696
8.	MARINE AND MARITIME USES	46 p.	N76-18769
9.	MATERIALS PROCESSING IN SPACE	40 p.	N76-18162
10.	INSTITUTIONAL ARRANGE-MENTS	56 p.	N76-18195
11.	COSTS AND BENEFITS	85 p.	N76-18995
12.	SPACE TRANSPORTATION	40 p.	N76-18199
13.	INFORMATION SERVICES AND INFORMATION PROCESSING	38 p.	N76-18988

66. National Academy of Sciences, Washington, D.C. USEFUL APPLICA-TIONS OF EARTH-ORIENTED SATELLITES. 15 vols. Prepared for NASA by the Summer Study on Space Applications personnel. Wash-ington, D.C.: National Academy of Sciences, National Research Council, 1969. Paper. Available from NTIS.

> In the fall of 1966, NASA asked NAS to conduct a study on "the probable future usefulness of satellites in practical Earth-oriented applications." Later, NASA requested that the study include considerations of economic factors. Work began in January 1967, and a series of thirteen technical panels were convened to study practical space applications and work was continued for two summers (1966, 1967) at Woods Hole, Massachusetts. Each panel provided reports which are listed below.

Panel 1.	FORESTRY--AGRICULTURE--GEOGRAPHY	76 p.	N69-27962
2.	GEOLOGY	73 p.	N69-28160
3.	HYDROLOGY	81 p.	N69-28360
4.	METEOROLOGY	83 p.	N69-28102
5.	OCEANOGRAPHY	117 p.	N69-28072
6.	SENSORS AND DATA SYSTEMS	94 p.	N69-27754
7.	POINTS-TO-POINTS COMMUN-ICATIONS		
8.	SYSTEMS FOR REMOTE-SENSING INFORMATION AND DISTRI-BUTION		
9.	POINT-TO-POINT COMMUNI-CATION	181 p.	N69-27876
10.	BROADCASTING	129 p.	N69-28938
11.	NAVIGATION AND TRAFFIC CONTROL	94 p.	N69-27963
12.	ECONOMIC ANALYSIS		
13.	GEODESY AND CARTOGRAPHY		
Report of Central Review Committee		40 p.	N69-27755
Summaries of Panel Reports		96 p.	N69-28240

67. National Research Council. Committee on Remote Sensing for Agricul-tural Purposes. REMOTE SENSING WITH SPECIAL REFERENCE TO

AGRICULTURE AND FORESTRY. Washington, D.C.: National Academy of Sciences, 1970. xiii, 424 p. Figs., illus., tables, maps.

This book consists of eight chapters and discusses the sensors, the analysis of the various data sets, and the applications and potential applications of these data for working on agricultural and forestry problems. This book is slightly dated, but is considered a classic. It was this publication, probably more than any other, that gave remote sensing of earth resources a major thrust into the prominence it presently has.

68. National Research Council. Committee on Remote Sensing Programs for Earth Resources Surveys. MICROWAVE REMOTE SENSING FROM SPACE FOR EARTH RESOURCE SURVEYS. Washington, D.C.: National Academy of Sciences, October 1977. ix, 139 p. Figs., tables, refs. Available from NTIS, N79-10497.

This report is a review, by the Committee on Remote Sensing Programs for Earth Resources Surveys of a NASA proposed MICROWAVE REMOTE SENSING PROGRAM FIVE YEAR TECHNICAL PLAN, 1977-1982. It is an especially interesting document because it presents a rather candid and possibly controversial review of the NASA proposals. Basically, the committee supports passive microwave for soil moisture and salinity studies and single polarization radars for geological studies. The arguments and data used by the committee are included in the report. Thus, this report provides quite interesting reading with respect to the decisions which are being made concerning microwave remote sensing from satellites.

69. National Research Council. Committee on Remote Sensing for Development. RESOURCE SENSING FROM SPACE: PROSPECTS FOR DEVELOPING COUNTRIES. Washington, D.C.: National Academy of Sciences, March 1977. 210 p. Tables, charts, maps, refs. Available from NTIS, PB-264-171.

In this report the committee discusses the matter of resource information and its use in developing countries. It then describes the new technology of earth remote sensing from space. In subsequent chapters a summary of tested usability of the satellite data in various sectors of application and the improvements anticipated from future technological advances are presented. The report then considers the prospects and problems of utilization of satellite data by developing countries and proceeds to the international issues that need to be resolved to facilitate widespread availability and exploitation of the data. The report concludes with recommendations that address certain policy issues and technical cooperation initiatives that would foster the diffusion of the technology.

70. National Scientific Laboratories. FREQUENCY REQUIREMENTS FOR ACTIVE EARTH OBSERVATION SENSORS. Falls Church, Va.: May 1977. 109 p. Appendixes. Available from NTIS.

> The purpose of this study, which was conducted under contract from the Jet Propulsion Laboratory (Contract 954669) is to present the foundation and rationale for the selection of microwave frequencies for active remote-sensing usage for subsequent use in the determination of sharing criteria and allocation strategies for the 1979 World Administrative Radio Conference. The study is basically a review of the literature from this point of view and, as such, it contains an extensive bibliography for active microwave sensing applications. The study consists of six appendixes, addressing each of the following areas: (1) soil moisture, (2) vegetation, (3) ice, (4) wave structures, (5) wind, and (6) rain measurements. A set of seven frequency allocations are determined to be needed if active microwave sensors are to achieve their earth-resources monitoring potential.

71. Nefedov, K.E., and Popova, T.A. DECIPHERING OF GROUNDWATER FROM AERIAL PHOTOGRAPHS [Deshifrirovanie gruntovykh vod po aerofotosnimkam]. Translated from the Russian by Dr. V.S. Kothekar for the National Aeronautics and Space Administration and the National Science Foundation. NASA Technical Translation F-681. Washington, D.C.: National Aeronautics and Space Administration, and the National Science Foundation, 1972. xii, 191 p. Figs., tables., refs. Available from NTIS, N73-11409.

> "The monograph is devoted to the use of aerial photographs in groundwater studies. The principles of groundwater photo interpretation, aerial photo sampling, and extrapolation of aerial photograph indexes are described. The technique is given for the medium-scale mapping of groundwater in areas of deficient precipitation. The benefits of this technique have been displayed in action. A number of landscape elements and morphological units are considered in terms of the estimation of groundwater conditions.
>
> The monograph can be recommended to hydrologists, geologists, specialists in melioration, and designers. It may serve as a manual to students of geology, hydrology, and meteorology both in higher and secondary schools." It provides an excellent background for this specialized aerial photograph interpretation.

72. Newhall, Beaumont. AIRBORNE CAMERA: THE WORLD FROM THE AIR AND OUTER SPACE. New York: Hastings House, 1969. 144 p. Illus.

> Since the first airborne camera in 1858, various systems and sensors have been used to view the earth from above. In

this short book the history of the development of such views
is briefly developed up to, and including, Gemini and Apollo
spacecraft photography. Some introductory remarks concerning multispectral sensing and radars are also included. This
is an excellent brief introduction to remote sensing with a
certain artistic flair, presenting some most unusual patterns
of natural features of the earth's surface.

73. Nicks, Oran W., ed. THIS ISLAND EARTH. NASA SP-250. Wash-
     ington, D.C.: Scientific and Technical Information Division, National
     Aeronautics and Space Administration, 1970. iv, 182 p. Col. photos.
     Available from U.S. Government Printing Office and NTIS, N71-12566.

     Photographs taken from satellites and manned spacecraft are
     presented and discussed in order that man may better under-
     stand his place in the universe. The emphasis is on the
     earth and its atmosphere, waters, lands, and man-made fea-
     tures. Detailed pictures are given of North American moun-
     tains, plains, and coasts. Photographs are also included of
     the sun, planets, and moon. This is an excellent book of
     color photography and imagery, and complements two other
     similar books in the NASA SP series, namely SP-129 (EARTH
     PHOTOGRAPHS FROM GEMINI III, IV, AND V, see cita-
     tion no. 119) and SP-168 (EXPLORING SPACE WITH A
     CAMERA, see citation no. 20).

74. Ordway, Frederick Ira III. PICTORIAL GUIDE TO PLANET EARTH.
     Pictoral Guides. New York: Thomas Y. Crowell Co., 1975. x,
     191 p. Illus., figs., tables.

     This book is one of a series prepared by this publisher and
     presents a brief history of the development of space imaging
     systems. Numerous examples of imagery obtained from Land-
     sat, Nimbus, Apollo, and various aircraft systems are in-
     cluded. This book has a good written guide to supplement
     the imagery.

75. Page, Robert N. THE ORIGIN OF RADAR. Science Study Series no.
     S-26. Garden City, N.Y.: Doubleday and Co., Anchor Books, 1962.
     196 p. Figs., illus. Paper.

     This small book provides a valuable background for the under-
     standing of radar by the lay person. The explanation of the
     principles and the concepts, together with their execution
     are clearly and concisely made. This publication deals pri-
     marily with military and navigation systems. Radar imagery,
     per se, is not addressed.

76. Pierce, John Robinson. ELECTRONS AND WAVES: AN INTRODUC-
     TION TO THE SCIENCE OF ELECTRONICS AND COMMUNICATION.

Science Study Series no. S-38. Garden City, N.Y.: Doubleday and Co., Anchor Books, 1964. xii, 226 p. Paper.

This revision of the first eight chapters of an earlier text presents a clear and lucid text of the fundamentals of electronics and EM waves. It also provides a valuable introductory statement for the nonscientists' understanding of some of the basic principles of remote sensing.

77. Porter, Ronald A., and Florance, Edwin T. FEASIBILITY STUDY OF MICROWAVE RADIOMETRIC REMOTE SENSING. 3 vols. Prepared for NASA under contract NAS 12-629. Cambridge, Mass.: January 1969. Available from NTIS, N70-14446, N70-14447, and N70-14448.

A comprehensive study on the feasibility of remote sensing by earth-orbiting microwave radiometers operating in the frequency range of 1-220 GHz was conducted. A broad range of terrestrial materials and phenomena were considered. Material emissivities were determined and incorporated in expressions designed to compute apparent or brightness temperatures and contrasts at a given spacecraft radiometer frequency. The effects of the atmosphere were taken into account and the state-of-the-art radiometer temperature sensitivities were determined and used as a basis for the signal-to-noise ratio. An analysis of the calculated contrasts revealed thirty-three pairs of material phenomena were detectable with reasonable certainty in range of weather conditions. The study showed a need for additional quantitative apparent temperature data and dielectric permittivity information relative to terrestrial materials. There is also a stated need for the development of techniques suitable for radiometric measurement of ocean roughness and temperature and for the interpretation of radiometric data. In summary, this work presents much of the needed background for understanding and interpreting the data obtained by the Nimbus series of satellites.

78. Pouquet, Jean. EARTH SCIENCES IN THE AGE OF THE SATELLITE [Les sciences de la terre a l'heure des satellites]. Translated from the French by I.A. Leonard. Boston: D. Reidel Publishing Co., 1974. vii, 169 p. Illus, figs., bibliog.

Following a brief review of the beginnings of remote sensing, the basic principles are outlined. It is considered that remote-sensing techniques will occupy first place in earth science studies. Types of satellites are described. Treatment of information received from satellites is studied, and comparative evaluation is made of aircraft versus satellite data. The use of manned satellites and general principles or interpretation are considered. Some geopedological and nongeopedological examples are discussed as well as prospects for the future.

79. Price, Alfred. INSTRUMENTS OF DARKNESS. London: William Kimber, 1967. 254 p. Illus., maps, figs., index.

> This excellent and easily read book gives the background of the development of radars during World War II and the following years. This account provides an interesting background for those interested in the history of remote sensing and again emphasizes the military background for many of the remote-sensing devices presently used for earth resources and other peaceful pursuits.

80. Ray, Richard G. AERIAL PHOTOGRAPHS IN GEOLOGIC INTERPRETATION AND MAPPING. U.S. Geological Survey Professional Paper 373. Washington, D.C.: U.S. Government Printing Office, 1960. vi, 230 p. Illus. Paper.

> This is an excellent publication dealing with the use of aerial photographs to obtain qualitative and quantitative geologic information as well as instrument procedures employed in compiling geologic data from aerial photographs. Many examples and well-reproduced stereo pairs are included.

81. Raytheon Co. Autometric Operation. DATA REDUCTION OF AIRBORNE SENSOR RECORDS. Report DOT-CG-01,800A. Alexandria, Va.: July 1970. 210 p. Figs., tables, refs.

> The capabilities of four remote sensors--panoramic camera, thermal infrared scanner, laser profiler, and side-looking airborne radar--were examined to determine the level of ice detail available from individual sensors and from the combination of sensors. A descriptive text of each sensor's ability to detect sea and fast ice types, conditions and phenomena are presented, along with photographic examples of ice types depicted on each sensor's record. An investigation was undertaken to establish data-base requirements for the storage, retrieval, and analysis of sea and fast ice data. This is an excellent example of how to approach multisensor analysis of a given scene.

82. Reifsnyder, William E., and Lull, Howard W. RADIANT ENERGY IN RELATION FOR FORESTS. USDA, Forest Service Technical Bulletin no. 1344. Washington, D.C.: U.S. Government Printing Office, December 1965. iv, 111 p.

> This technical bulletin reviews, for foresters and those interested in the forest complex, the fundamentals of short- and long-wave radiation, of radiation measurement, and of certain radiation relationships in the forest. The manner in which the forest absorbs, reflects, radiates, and transmits radiant energy is also presented. As such, this booklet provides quite a solid understanding for the use of remote sensing in the study of forest and other vegetative ecosystems.

83. Reintjes, J. Francis, and Coate, Godfrey T., eds. PRINCIPLES OF RADAR. 3d ed. New York: McGraw-Hill Book Co., 1952. iv, 985 p. Figs., tables, index.

> This book consists of fourteen chapters dealing with the various engineering designs and principles of radars. Military type, nonimaging radars are the major concern. No comments concerning earth resources are included, but the publication provides a good introductory statement of radar principles. Synthetic aperture radars which are normally used for earth resources are not included.

84. Rinker, Jack N., et al. CAPABILITIES OF REMOTE SENSORS TO DETERMINE ENVIRONMENTAL INFORMATION FOR COMBAT. ETL-0081. Fort Belvoir, Va.: U.S. Army Engineer Topographic Laboratories, November 1976. 245 p. Available from NTIS, N77-25615.

> U.S. Army field and technical manuals were used to develop a list of 313 environmental information needs required by the U.S. Army to accomplish its various tasks. Each factor was evaluated against a list of remote-sensing systems which include Landsat, radar, thermal infrared, low-level oblique photography, standard photo index sheets, stereo 1:100,000 vertical aerial photography, and stereo 1:20,000 vertical aerial photography. Interpretation procedures were restricted to evaluation of imagery by conventional interpretation techniques and equipment. The level of the interpretation skills needed and the capability of the remote sensor to detect each of the environmental factors are categorized. This paper is an excellent source for a broad comparative evaluation of the various remote sensors which are in common use in the civilian community.

85. Robinove, Charles J. WORLDWIDE DISASTER WARNING AND ASSESSMENT WITH EARTH RESOURCES TECHNOLOGY SATELLITES. Project Report (IR) NC-47. Washington, D.C.: U.S. Geological Survey, 1975. ii, 65 p. Figs., tables.

> The purpose of this report, prepared for the Disaster Coordination Office of the Agency for International Development by USGS is to define, on the basis of experimental results, the potential use of Earth Resources Technology Satellites for worldwide disaster monitoring, techniques, and problems. The report is not exhaustive but does contain a series of abstracts describing results of experiments to study floods, earthquakes, volcanic eruptions, droughts, grass and forest fires, agricultural crop disasters, glacier movement, and water pollution monitoring.

86. Romanova, Mariya A. AIR SURVEY OF SAND DEPOSITS BY SPECTRAL LUMINANCE [Opredelenie tipovogo sostava peschanykh otlozhenii s

vozdukha po ikh spektral'noi yarkosti]. Translated from the Russian by Consultants Bureau, New York. New York: Consultants Bureau, 1964. 158 p. Figs., tables, refs., appendixes.

This book gives an account of a method for determining the composition of sandy deposits from the air by the nature of their reflections of radiant energy. The measured spectral luminance of a rock outcrop relative to the luminance of a standard is the basic property considered in this method. The text consists of seven chapters covering the basic concepts of photometry, methods of measuring spectral luminance, evaluation of the data, and examples of the methods used. This authorized translation from the Russian contains revisions and supplementary material provided by the author.

87. Rosenfeld, Azriel. PICTURE PROCESSING BY COMPUTER. New York: Academic Press, 1969. x, 196 p. Figs., illus.

The author assumes a background in calculus in this review of the concepts and techniques of computer picture processing. It contains extensive references as well as author and subject indexes.

88. Rudd, Robert D. REMOTE SENSING: A BETTER VIEW. The Man-Environment System in the Late Twentieth Century, edited by William Thomas. North Scituate, Mass.: Duxbury Press, 1974. 135 p. Plates, illus., refs., index.

The purpose of this brief volume is to introduce more people to remote sensing and to make them think about its potential implicit in the pioneering efforts in research and application that have been carried out to date. Numerous examples of applications of remote sensing to earth-resources studies and the basic introductory text to the nature of remote sensors amply succeed in achieving the stated purpose.

89. Ruechardt, Eduard. LIGHT: VISIBLE AND INVISIBLE. Translated from the German by Frank Gaynor. Ann Arbor: UNIVERSITY OF MICHIGAN PRESS, 1958. 201 p.

This introductory and nonmathematical book introduces, explains, and discusses the nature of EM waves, including diffraction, interference, polarization and double refraction, and scattering. An excellent and valuable background for nonphysicists and nonengineers for providing an understanding of remote-sensing instrumentation and conceptual theory is presented in this publication.

90. Rydstrom, Hubert O. GEOLOGIC EXPLORATION WITH HIGH RESO-LUTION RADAR. GIB-9193A. Litchfield Park, Ariz.: Goodyear Aerospace Corp., 1970. vi, 48 p. Figs., illus. Paper, spiralbound.

A number of natural features observed by high resolution
side-looking radar are annotated and briefly analyzed.
Another group of images is analyzed in greater detail and
the results applied to specific geologic problems in such
fields as petroleum, groundwater and metallic minerals. The
introduction briefly discusses and illustrates some of the principles of imaging radars as they apply to interpretation of
images of earth resources.

91.   Rydstrom, Hubert O.; LaPrade, G.L.; and Leonardo, E.S. MILITARY
       THEMATIC MAPPING AND MAP COMPILATION FROM RADAR IM-
       AGERY. Prepared for U.S. Army Engineer Topographic Laboratories
       under contract no. DAAKO2-72-C-0508. Litchfield Park, Ariz.:
       Goodyear Aerospace Corp., July 1973. viii, 228 p. Figs., refs.

       This report, designed primarily as a course for the teaching
       of radar interpretation techniques, is one of the most com-
       plete and knowledgeable references on the subject. Al-
       though dealing primarily with X-Band interpretation tech-
       niques, the principles and examples are easily (and generally)
       extended to all other wavelengths of active microwave im-
       aging sensors. Subjects discussed include, in addition to
       the functioning of the radar and the processing equipment,
       the earth science disciplines of vegetation, rocks and soils,
       topography, drainage, hydrography, and some cultural fea-
       tures such as lines of communications and urban areas.

91A   Sabins, Floyd F., Jr. REMOTE SENSING: PRINCIPLES AND INTER-
       PRETATION. San Francisco: W.H. Freeman and Co. xi, 426 p.
       Figs., tables, illus., index, glossary, maps.

       This book, written by an individual who has vast and wide
       practical experience in the field of remote sensing, covers
       the gammet of remote sensing from the fundamental consider-
       ations to applications in specific fields. The strength of
       this book, which was designed for a one-semester course for
       upper-class or graduate students, lies primarily in the fields
       of geologic applications and associated topics. Clearly, this
       is one of the strongest remote sensing books to be presently
       available for the earth science community.

92.   Saint Joseph, K.S., ed. THE USES OF AIR PHOTOGRAPHY. 2d ed.
       London: John Baker, 1977. 194 p. Refs., tables, figs.

       This book considers the scope of aerial photography, includ-
       ing cartography based on aerial photography, the instruments
       and techniques used in air photo interpretation, and a con-
       sideration of applications in a wide array of earth sciences.

93.   Scherz, James P., and Stevens, Alan R. AN INTRODUCTION TO

REMOTE SENSING FOR ENVIRONMENTAL MONITORING. Edited by Edmund R. Belak, Jr. Project supported by NASA Multidisciplinary Research Grant in Space Science and Engineering. Institute for Environmental Studies, Engineering Experiment Station, Graduate School, Report no. 1. Madison: Remote Sensing Program, University of Wisconsin, August 1970. 80 p. Figs., tables. Paper.

This report gives an overall view of the electromagnetic spectrum. It describes remote-sensing instruments that have potential use from aircraft and/or spacecraft. Emphasis is placed on photographic, thermal imagery, side-looking radar, and passive microwave systems. The report consists of four major parts: (1) theory of EM spectrum, (2) hardware presently used in remote sensing, (3) present and future applications of remote sensing, and (4) sources of remote-sensing material.

94. Shepard, Francis P., and Wanless, Harold R. OUR CHANGING COASTLINES. New York: McGraw-Hill Book Co., 1971. 579 p. Illus., tables, refs.

Although the thrust of this publication is to show the changes in the coastline of the United States, this book is an important source of aerial photography information for those interested in coastal processes. The entire coast of the United States, including Alaska and Hawaii, is discussed and numerous photographs, together with discussion of the use of these for geomorphological and geological studies, are included.

95. Short, Nicholas S., et al. MISSION TO EARTH: LANDSAT VIEWS THE WORLD. NASA SP-360. Washington, D.C.: National Aeronautics and Space Administration, U.S. 1976. ix, 459 p. Illus., figs., glossary. Available from the U.S. Government Printing Office.

This excellent book, which may be characterized as a space photo album, contains a very comprehensive guide to the Landsat satellite and its products. The first section consisting of twenty-seven pages, considers the Landsat system, the instrumentation, the data, and some applications of these data. This is followed by a selection of approximately four hundred false color Landsat images with accompanying interpretative and application comments. A glossary of technical terms and an index to the several plates comprise the appendix of this publication.

96. Simonett, David S., ed. APPLICATIONS REVIEW FOR A SPACE PROGRAM IMAGING RADAR. Geography Remote Sensing Unit Technical Report no. 1. Prepared under Contract NAS 9-14816 from NASA Johnson Space Center. Santa Barbara: Department of Geography, University of California, July 1976. 231 p. Illus., bibliog., figs.

Paper. Available from NTIS, N77-18309.

This report consists primarily of an expanded and modified briefing which was presented to NASA headquarters and prepared by the Space Program Imaging Radar Group composed of scientists and engineers from government and universities. The report consists of nine chapters: (1) "Introduction and Executive Summary," (2) "Microwave Sensing of Water Resources," (3) "Space Shuttle Imaging Radar Roles in Mineral and Petroleum Exploration," (4) "Vegetation Resources Analysis with Radar," (5) "Space Radar Applications for Ocean and Ship Monitoring," (6) "Space Shuttle Imaging Radar and Its Applicability to Mapping, Charting, and Geodesy," (7) "State Applications of a Space Program Imaging Radar," (8) "Federal Agency Requirements," with an example, and (9) "Some Roles of Shuttle Radar." It is quite a comprehensive review paper.

97. Singh, R. "A Practical Cataloging, Indexing, and Retrieval System for Remote Sensing Data." Ph.D. Dissertation, University of Wisconsin, 1973. xiv, 148 p. Tables, figs.

An introductory statement concerning cataloging data, designed primarily for use with low-altitude aerial photography but also capable of being used for virtually any remotely sensed data, is presented. The system consists of a combination of standard book catalog systems and has elements of many map cataloging systems. Because one is dealing with both scale, location, subject, and type of imagery, the problem of cataloging can become quite complex. In that context, this publication provides useful guidance.

98. Skolnik, Merrill I., ed. in chief. RADAR HANDBOOK. McGraw-Hill Handbook Series. New York: McGraw-Hill Book Co., 1970. ix, 1,909 p. Illus., figs., refs., index.

This publication is an exceptionally useful publication for use with radar and radar remote sensing. It contains a total of thirty-six chapters, many of which are devoted to the engineering and technological aspects of all types of radars. Chapter 1 (by M.I. Skolnick, "An Introduction to Radar") and chapter 25 (by R.K. Moore, "Ground Echo") are especially useful for earth resources applications of SAR and SLAR remote-sensing systems.

98A Smith, H.T.U. PHOTO-INTERPRETATION STUDIES OF DESERT BASINS IN NORTHERN AFRICA. Report AFCRL, 68-0590. Bedford, Mass.: Air Force Cambridge Research Laboratories, June 1969. 77 p. Illus., figs.

The data used in this study were a series of medium-scale

Tri-metrogon aerial photos. The text considers the regional setting with respect to geography, geology and climate, and then proceeds to the interpretation of a series of photographs. This is one of the better guides to the aerial photograph interpretation of desert regions, in that it points out significant differences between the northern Africa desert terrain and that found in the western United States.

99. Smith, William L., ed. REMOTE-SENSING APPLICATIONS FOR MINERAL EXPLORATION. Stroudsburg, Pa.: Dowden, Hutchinson and Ross, 1977. xiv, 391 p. Tables, refs., figs., index.

The purpose of this book is to take a broad look at the early returns from a new technology as they relate to mineral exploration and foreseeable problems in mineral resources management. Thus, it is an attempt to synthesize new concepts and capabilities that have been gained largely in the past few years since Landsat was launched. The book consists of fifteen chapters dealing with mineral resources problems and the ways that remotely sensed data, its processing, and interpretation, may aid in solving the identified problems.

100. Spurr, Stephen H. PHOTOGRAMMETRY AND PHOTO-INTERPRETATION: WITH A SECTION ON APPLICATIONS TO FORESTRY. New York: Ronald Press Co., 1960. vi, 472 p. Figs., plates, charts.

This is the second edition (formerly titled AERIAL PHOTOGRAPHS IN FORESTRY) which has been extensively rewritten to cover other earth science disciplines. Although somewhat dated with respect to the instrumentation discussed and pictured, the comments concerning photogrammetry, mapping, aerial photography, and photo interpretation remain valid and quite useful. Still one of the better sources for photo interpretation in the area of silviculture.

101. Stanford University. School of Engineering. DEMETER: AN EARTH RESOURCES SATELLITE SYSTEM. Stanford, Calif.: June 1968. 526 p. Paper. Available from NTIS, N69-35000.

The Demeter satellite system would observe the earth in the optical, near infrared, and thermal infrared wavelengths and produce real time multiband pictures approximately every two weeks for distribution to the various users. Studies presented in this report indicate annual benefits on the order of billions of dollars which make the six satellites and the four-year costs of $103 million well justified. The study also presents the economic and political as well as technical points of view. The technical studies are based on an operational date of 1973 and anticipate most advancements of the state-of-the-art. They pursue the design of new systems through most of the theoretical problems to a point where production seems feasible. This report is the result of a class project in engineering in which seventy-three students participated. The report illustrates some of the thoughts and considerations that go into designing an earth-resources satellite system.

102. Tarkoy, Peter J. INTRODUCTION TO BASIC REMOTE SENSING FOR ENGINEERING GEOLOGISTS. Urbana: Rock Mechanics Laboratory, University of Illinois, June 1971. 169 p. Paper. Available from NTIS, PB-224-754.

The objective of the report is to present a technical discussion of the theory of the electromagnetic spectrum and relevance to remote sensors, describe the bands and sensors used in applications, and relate how they can be useful in engineering geology and geological exploration. A discussion of each sensor's capabilities, most useful application, and examples of past experiences are also included. Sections on electromagnetic radiation, photographic systems, data procurement, nonphotographic systems, and geologic applications of remote-sensing data are included along with a short annotated bibliography and an extensive review of articles dealing with applications of remote sensing in engineering geology.

103. TRW Systems Group. EARTH RESOURCES MISSION PERFORMANCE STUDIES. 2 vols. Prepared under NASA contract NAS9-14117 for L.B. Johnson Space Center, Houston, Tex. Redondo Beach, Calif.: August 1974. 2 p., 68 p. Tables, figs. Available from NTIS, N76-19537 and N76-19538.

The requirements definition task (covered by the first volume) for the Earth Resources Mission Performance Study was limited to an update and consolidation of information presently available. The prime objective of this task was to gather the diverse user agency requirements for remote-sensing data and to organize them into sensor collection requirements for and EOS-A configuration including a thematic mapper (TM) and a high resolution pointable imager (HRPI). The second volume has as its purpose the analysis of the results developed from a simulation of an EOS mission. Two sensors were considered, the TM and HRPI. Simulations included the earth resources applications requirements for the contiguous U.S. land and coastal areas only. These two volumes are a good source for information concerning satellite project planning.

104. TRW Systems Group. EARTH RESOURCES TECHNOLOGY SATELLITE. 18 vols. Prepared for NASA under contract NAS5-11250. Redondo Beach, Calif.: 1970. Available from NTIS.

Because of the importance of the ERTS (now Landsat) satellite for earth resources information gathering and for earth scientists in general, questions concerning the nature of the satellite, its operation, the system, and its engineering design and functioning are often asked. This series of reports provide one of the most accessible sources of information for answering such questions.

CONTENTS

Vol. 1. SUMMARY   21 April 1970
         111 p.   N70-34409

Vol. 2. SYSTEM STUDIES   17 April 1970
         281 p.   N70-34410

Vol. 3. OBSERVATORY SYSTEM DESIGN   11 February 1970
         265 p.   N70-34411

Vol. 4. OBSERVATORY SUBSYSTEMS STUDY   11 February 1970
         529 p.   N70-34412

Vol. 5. DATA COLLECTION SYSTEM   11 February 1970
         265 p.   N70-34413

Vol. 6. RELIABILITY PROGRAM PLAN   17 April 1970
         88 p.   N70-34414

Vol. 7. QUALITY PROGRAM PLAN   17 April 1970
         172 p.   N70-34415

Vol. 8. TEST MONITORING AND CONTROL   17 April 1970
         PLAN
         34 p.   N70-34416

Vol. 9. CONFIGURATION MANAGEMENT   17 April 1970
         PLAN
         114 p.   N70-34417

Vol. 10. SOLDERING PROGRAM PLAN   11 February 1970
         70 p.   N70-34418

Vol. 11. FAILURE REPORTING PLAN   17 April 1970
         35 p.   N70-34419

Vol. 12. OBSERVATORY INTEGRATION AND   11 February 1970
         TEST PLAN AND LAUNCH
         OPERATIONS PLAN
         81 p.   N70-34420

Vol. 13. GROUND DATA HANDLING SYSTEM   17 April 1970
         STAFFING AND MATERIAL USAGE
         PLAN
         79 p.   N70-34421

Vol. 14.   GROUND DATA HANDLING SYSTEM   17 April 1970
           DESIGN
              384 p.   N70-34422

           APPENDIX. OPERATIONS CONTROL   17 April 1970
           CENTER AND NATIONAL DATA
           PROCESSING FACILITY SPECIFICATIONS
              165 p.   N70-34423

Vol. 15.   GROUND DATA HANDLING SYSTEM   17 April 1970
           STUDY
              71 p.   N70-34424

           APPENDIX. FUNCTIONAL ANALYSIS   17 April 1970
              168 p.   N70-34425

Vol. 16.   OPERATIONS CONTROL CENTER   17 April 1970
           STUDIES
              211 p.   N70-34426

Vol. 17.   NASA DATA PROCESSING FACILITY   17 April 1970
           STUDIES
              401 p.   N70-34427

Vol. 18.   AUTOMATIC DATA PROCESSING   17 April 1970
           EQUIPMENT PROCUREMENT
              37 p.   N70-34428

104A. Twomey, C. INTRODUCTION TO THE MATHEMATICS OF INVERSION
IN REMOTE SENSING AND INDIRECT MEASUREMENTS. New York:
Elsevier Scientific Publishing Co., 1977. 242 p.

> Although the word "introduction" appears in the title, this
> publication is apparently a very sophisticated presentation
> of the subject under discussion, and is appropriate only to
> those who have a relatively secure mathematical background.
> See the review by A.J. LaRocca in REMOTE SENSING OF
> ENVIRONMENT 4 (October 1978): 363-65.

105.   U.S. Army. Corps of Engineers. EARTH RESOURCES SURVEYS FROM
SPACECRAFT. 2 vols. Washington, D.C.: Space Applications Pro-
grams Office, Earth Resources Survey Program, NASA, n.d. Va. pag.
Looseleaf.

> The purpose of this document is to acquaint the scientific
> community and interested public with some of the potential
> applications of the space program. These volumes illustrate

some of the applications of space for the earth-resources
disciplines of agriculture, forestry, geology, hydrology,
oceanography, geography, and cartography. The results of
interpretation of selected examples relating to a variety of
geoscience topics are included. These samples serve to
show the current state of progress in obtaining and analyzing
synoptic view imagery. These two volumes show numerous
excellent examples of Gemini and Apollo (manned) spacecraft photography, complete with interpretations and sketch
maps. Also included are some descriptions of various remote-
sensing instruments and brief discussions of the interpretation
methodology. Excellent, but very difficult to locate as a
limited number were published, and all apparently distributed
on permanent loan to various institutions and investigators.

106. U.S. Congress. House of Representatives. Committee on Science
and Astronautics. EARTH RESOURCES SURVEY SYSTEM. HEARINGS
ON H.R. 14978 AND H.R. 15781 BEFORE THE SUBCOMMITTEE ON
SPACE SCIENCE AND APPLICATIONS, 93d Cong., 2d sess., 1974.
Washington, D.C.: U.S. Government Printing Office, 1974. 286 p.
Also available from NTIS, N75-12416.

These hearings explore the desirability of creating an operational satellite system for surveying the earth's resources.
The ERTS program is reviewed and the technology for data
acquisition, dissemination, and utilization are assessed along
with the technology of remote sensing. These hearings were
held between 3 and 9 October 1974.

107. U.S. Congress. House of Representatives. Committee on Science and
Astronautics. REMOTE SENSING OF EARTH RESOURCES. A Compi-
lation of Papers prepared for the 13th Meeting of the Panel on Science
and Technology--1972. Washington, D.C.: U.S. Government Printing
Office, 1972. ix, 224 p. Illus. Paper. Also available from NTIS,
N72-20958, and N72-23307.

Testimony given before the Committee on Science and Astronautics on the current state of the technology applicable to
the research and development phase of remote-sensing systems
of earth resources is presented. These presentations include
the following: (1) projections of additional requirements needed
as these systems approach operational status, (2) identification
of potential users of various data to be produced by the systems, (3) organizational framework needed to establish effective
operational systems, (4) international implications of remote-
sensing systems, and (5) problems of data handling and dissemination.

108. U.S. Congress. Senate. Committee on Aeronautical and Space Sciences. EARTH RESOURCES SATELLITES. HEARINGS ON S. 2350
AND S. 3484 BEFORE THE COMMITTEE ON AERONAUTICAL AND

SPACE SCIENCES UNITED STATES SENATE, 93d Cong., 2d sess., 1974. Washington, D.C.: U.S. Government Printing Office, 1974. 341 p. Available from NTIS, N75-13010.

This publication contains hearings presented before the U.S. Senate on 6, 8, and 9 August 1974 and on 18 September 1974. The arguments for and against the expansion of earth-resources satellite capabilities and preparation for an operational satellite system were presented in a hearing before Congress, whose purpose was to gather information on two impending bills which would increase the federal government's activities in remote-sensing programs. Significant accomplishments of ERTS-1 are reviewed, along with some potential capabilities which are as yet unexplored. The ways in which ERTS data are being used are also described with emphasis on the role of the EROS Data Center in Sioux Falls, South Dakota, in disseminating information of use to various individuals and agencies. Plans for an operational earth-resources satellite system, which is expected to become a more pressing issue with the launch of ERTS-B, are discussed in terms of public vs. private control and the possibility that such a system will present difficulties in foreign relations.

109. U.S. National Aeronautics and Space Administration. EARTH RE-SOURCES TECHNOLOGY SATELLITE A: PRESS KIT. Washington, D.C.: July 1972. 99 p. Illus. Available from NTIS, N72-26811.

This is an interesting item, in that it is the complete press kit, to be released on 20 July 1972, the day before the projected launch of ERTS-A. It provides a wealth of information concerning the satellite system and the experiments which were to be conducted. Written for the press to use, almost without modification, and easily understood by laymen.

110. U.S. National Aeronautics and Space Administration. Educational Programs Division. OBSERVING EARTH FROM SKYLAB. NASA Facts NF-56/1-75. Washington, D.C.: U.S. Government Printing Office, 1975. 16 p. Illus., maps, figs., bibliog.

This excellent small booklet provides background material on general remote sensing. Included are a map showing the extent of coverage of the earth from Skylab together with a description of the sensors and the nature of the data collected by Skylab. Also included is information on how to obtain earth-resources data from government agencies.

111. _____ . SKYLAB: A PREVIEW OF AMERICA'S FIRST EARTH-ORBIT-ING SPACE STATION. NASA Facts NF-43/1-72. Washington, D.C.: U.S. Government Printing Office, 1972. 1 sheet. 120 x 53 cm. Color.

This is a large wall sheet which briefly describes and illustrates the basic operation and mission objectives of the Skylab missions. A large 'cut-away' drawing of Skylab is presented.

112. _____. SPACE SHUTTLE. NASA Facts NF-44/7-72. Washington, D.C.: U.S. Government Printing Office, 1972. 1 sheet. 120 x 53 cm. Color.

This is a large wall sheet illustrating the concept, operation, and objectives of the Space Shuttle missions. Some of the objectives deal directly with the remote sensing of the earth environment.

113. _____. THE SPECTRUM: THERE'S MORE THAN MEETS THE EYE. NASA Facts NF-54/1-75. Washington, D.C.: U.S. Government Printing Office, 1975. 1 sheet. 120 x 53 cm. Color.

This is a four-color wall sheet with text and illustrations covering the electromagnetic spectrum which includes gamma rays, X-rays, ultraviolet light, visible light, infrared, and radio waves. It shows the relationship of these various wavelengths to one another and is an excellent tutorial aid.

114. _____. WHY SURVEY FROM SPACE? NASA Facts NF-57/1-75. Washington, D.C.: U.S. Government Printing Office, 1975. 9 p. Illus., figs.

This brochure provides some background of the nature and uses of aerial and space surveys of the earth. Good general introduction, especially useful for junior high school students.

115. U.S. National Aeronautics and Space Administration. Johnson Space Center. EARTH RESOURCES PROGRAM SYNOPSIS OF ACTIVITY. Houston: March 1970. 230 p. Figs., illus., charts. Paper.

According to the foreword of this book, during recent months, program reviews and discussion with interested groups have resulted in a number of charts which depict activities in the program. A selection of these charts has been combined into this document with a minimum of descriptive narrative. The purpose of the document is to provide general orientation material for individuals not intimately associated with the program. It is not intended to represent a complete program description or report but rather a cross section of the types of activity within the program at the Manned Spacecraft Center.

As such, this document consists of numerous charts, diagrams, and space photographs, together with limited narrative statements, concerning the instruments, platforms, and data which

were collected under the Earth Resources Program. An excellent document for introductory statements and examples
of these efforts.

116. _____. SKYLAB EXPLORES THE EARTH. NASA SP-380. Washington, D.C.: U.S. Government Printing Office, 1977. xii, 517 p.
Illus., figs., refs., appendixes, glossary, index.

This publication is concerned with the photographic earth
observations conducted on the four Skylab missions: (1) 14
May 1973, (2) 25 May 1973-22 June 1973, (3) 28 July
1973-25 September 1973, and (4) 16 November 1973-4 February 1974. A total of eighteen chapters, each dealing
with a specific problem of earth observations (e.g., snow
mapping, desert sand seas, vegetation patterns, mesoscale
cloud features, etc.) and three appendixes (glossary, standard weather symbols, and photographic index) comprise the
volume. Excellent color plate reproduction with annotations
as well as discussions and conclusions of the several Skylab
experiments are presented.

117. _____. SPACE SHUTTLE. NASA SP-407. Washington, D.C.: U.S.
Government Printing Office, 1976. viii, 89 p. Paper.

In the 1980s, Space Shuttle, a reusable vehicle for attain-
ing space satellite altitudes and orbiting the earth, will be
launched for scientific missions and earth-resources data
gathering. This small booklet describes in detail, the nature
of these missions, the equipment and instrumentation which
will be on board, the economic impact of such efforts, and
the nature of the shuttle craft itself. An excellent small
booklet.

118. U.S. National Aeronautics and Space Administration. Scientific and
Technical Information Division. EARTH PHOTOGRAPHS FROM GEMINI
VI THROUGH XII. NASA SP-171. Washington, D.C.: U.S. Gov-
ernment Printing Office, 1968. 335 p. Illus., refs., glossary.

The photographs in this volume resulted from two of the
twenty-two scientific experiments that were part of the Gemini
program between 15 December 1965 and 15 November 1966.
These excellently reproduced photographs are arranged geo-
graphically in nine sections. A complete listing of all photographs taken (in addition to those presented in this publication) is given in appendix B of the volume.

119. _____. EARTH PHOTOGRAPHS FROM GEMINI III, IV, AND V.
NASA SP-129. Washington, D.C.: U.S. Government Printing
Office, 1967. ix, 266 p. Illus., refs.

A collection of 244 color photographs obtained from Gemini 3, 4, and 5 are presented. Most of the pictures were obtained by Gemini 4 and 5 in a series of synoptic weather and synoptic terrain photography experiments, formally scheduled to obtain high-quality color photographs of terrain features and cloud systems for geological, geographical, and meteorological purposes. On the Gemini 4 flight priority was given to photography of east Africa, the Arabian Peninsula, Mexico, and the southwestern United States. For Gemini 5 the selected land and near-shore areas were chosen not only for geologic study, but also for geographic and oceanographic investigations. Although earth science photographs were not formally scheduled on Gemini 3, astronauts took twenty-five pictures, most of which showed cloud formations. Photographs are arranged in orbital sequence. Details concerning length of flight, apogee and perigee, crew, number of photographs taken, and photographic equipment used, preface each section. Captions accompanying the pictures identify political subdivisions, geographic features, and in some cases discuss geological and meteorological features.

120. _____. ECOLOGICAL SURVEYS FROM SPACE. NASA SP-230. Prepared for the Office of Space Science and Applications. Washington, D.C.: U.S. Government Printing Office, 1970. iv, 75 p. Illus., figs., refs.

Published prior to the launch of Landsat (originally ERTS) in 1972, this booklet presents applications within the ecological realm of Gemini and Apollo (manned) spacecraft photography. Subjects covered include geography, agriculture, forestry, geology, hydrology, oceanography, and cartography. Excellent photography together with sketches, maps, and discussion make this a very useful book for introductory statements about orbital remote-sensing applications.

121. _____. REMOTE MEASUREMENT OF POLLUTION. NASA SP-285. Washington, D.C.: U.S. Government Printing Office, 1971. vi, 253 p. Available from NTIS, N72-18324.

This publication is the report of a working group sponsored by NASA and convened at Norfolk, Va., 16–20 August 1971. Five subgroups (information on remote sensing, gaseous air pollution, water pollution, particles, and instrumentation) were formed. A summary of the major conclusions and recommendations developed by the five subgroups is presented. It becomes evident that many of the trace gases are amenable to remote sensing; that certain water pollutants can be measured by remote techniques, but that their number is limited, however; and that a similar approach to the remote measurement of specific particulate pollutants will follow only after understanding of their physical, chemical, and radiative properties is improved. It is also clear that remote sensing can provide essential information in all three categories that can not be obtained by any other means.

122. _____. A SURVEY OF SPACE APPLICATIONS. NASA SP-142. Washington, D.C.: U.S. Government Printing Office, April 1967. v, 135 p. Available from NTIS, N67-23338.

This document is a basic contribution of NASA to the 1967 study of space applications. The purpose of this survey is to focus attention on the real and potential applications of space technology; to summarize accomplished work, present efforts, and future plans; to identify policy and technical problems; and to provide a selected bibliography. Among other items covered are several earth resource areas (agriculture and forestry, geology and minerals, geography, cartography and cultural resources, hydrology and water resources, and oceanography).

123. _____. NASA THESAURUS. 2 vols. NASA SP-7050. Washington, D.C.: January 1976. Available from NTIS.

Volume 1, ALPHABETICAL LISTING, contains all subject terms (postable and nonpostable) approved for use in the NASA scientific and technical information system. Included are the subject terms of the preliminary edition of the NASA THESAURUS (NASA SP-7030, December 1967); of the NASA THESAURUS ALPHABETICAL UPDATE (NASA SP-7040, September 1971); and terms approved added or changed through 31 May, 1975. Thesaurus structuring, including scope notes, a generic structure with broader and narrower term (BT-NT) relationships displayed in embedded hierarchies, and other cross-references, is provided for each term, as appropriate. Volume 2, ACCESS VOCABULARY, contains an alphabetical listing of all thesaurus terms (postable and nonpostable) and permutations of all multiword and pseudomultiword terms. Also included are "Other Words" (nonthesaurus terms) consisting of abbreviations, chemical symbols, and so forth. The permutations and "Other Words" provide "access" to the appropriate postable entries in this thesaurus.

124. U.S. National Aeronautics and Space Administration. Skylab Program and Educational Programs Division. REMOTE SENSING OF EARTH RESOURCES. Skylab Experiments, Vol. 2. NASA EP-111. Washington, D.C.: U.S. Government Printing Office, May 1973. 88 p. Illus., figs. Paper.

This is the second of a seven-volume set (and the only one dealing directly with remote sensing) which describes all of the major aspects of remote sensing. Included are descriptions

of the individual earth-resources sensors and experiments included on Skylab.

125. U.S. Navy. Bureau of Ships. RADAR ELECTRONIC FUNDAMEN-
      TALS. NAVYSHIPS 900,016. Washington, D.C.: U.S. Government
      Printing Office, June 1944. v, 474 p.

      Although originally written for the "purpose of providing
      the student technician with an understanding of basic cir-
      cuits which are the components of more complex radar cir-
      cuits," this booklet provides helpful information for the
      earth scientists wishing to use radars for earth resources.
      This is accomplished by presenting basic concepts of radar
      engineering and thus providing a framework in which the
      earth scientist may operate vis-a-vis radar remote sensing.
      Written in 1944 and superseding TM 11-466, the booklet
      was available over the counter at U.S. Government Printing
      Office bookstores as late as December 1973, but is no longer
      in print.

126. Van Genderen, John L., and Collins, W. Gordan, eds. REMOTE
      SENSING DATA PROCESSING. Sheffield, Engl.: Remote Sensing
      Society, 1975. 148 p. Figs., illus., refs. Paper.

      This is a companion volume to FUNDAMENTALS OF REMOTE
      SENSING (see Collins and Van Genderen, item no. 154)
      which discusses a wide range of techniques for processing
      remote-sensing data. Visual, photographic, electronic, and
      automated machine processing are considered. Although lo-
      cal examples are considered, the thrust is on the principles
      and procedures of data handling, interpretation, and analy-
      sis which could be applied to the problems of earth-resources
      remote-sensing data applications.

126A Verstappen, Herman Theodoor. REMOTE SENSING IN GEOMORPH-
      OLOGY. Amsterdam: Elsevier Scientific Publishing Co., 1977. ix,
      214 p. Illus., figs., refs, tables.

      Professor Verstappen, a member of the faculty at the Inter-
      national Institute for Aerial Survey and Earth Science has
      produced a remarkably readable book dealing with all as-
      pects of remote sensing and the science of geomorphology.
      Excellent descriptions of the remote sensing techniques, de-
      tailed presentation of the characteristics of the remote sens-
      ing systems (specifically aerial photography, thermal imagery,
      and side looking airborne radar) are presented in an early
      chapter. The general principles of geomorphological image
      interpretation are presented and followed by separate chap-
      ters dealing specifically with geomorphological image analy-
      sis based on relief criteria and on density criteria. These

chapters are followed by several dealing with environmental geomorphology, genetic geomorphology, and several examples of geomorphological image interpretation. The publication is published in a most supurb manner, has excellent illustrations (both continuous tone and black and white) and includes extensive references. This is, to date, possibly one of the major publications in remote sensing and earth science.

127. Vinogradov, B.V., ed. AERIAL PHOTOGRAPHY USED IN MAPPING VEGETATION AND SOILS [Opyt kartirovaniya rastitel 'nosti i pochv po aerosnimkam]. Translated by Aerospace Technology Division, Library of Congress. Translations of Soviet-Block Scientific and Technical Literature. Washington, D.C.: U.S. Library of Congress, May 1965. 143 p. Refs. Available from NTIS, N65-29925 and AD 465 778.

A total of ten chapters dealing only with photo interpretation and application (i.e., no photogrammetry) are included in this translation. The entire collection describes the experience of using aerial methods in geobotanical and soil mapping under various natural conditions of the forest, steppe, semidesert and desert zones, and also in marine landscapes. The articles reflect the specific features of using aerial methods in solving various practical questions such as lumbering, botanical-fodder and soil reclamation. These chapters reflect the basic directions in the research being carried out (in 1964) by the Aerial Methods Laboratory for mapping vegetation and soil resources.

128. Vogel, Theodore C., et al. A MATRIX EVALUATION OF REMOTE SENSOR CAPABILITIES FOR MILITARY GEOGRAPHIC INFORMATION. Ft. Belvoir, Va.: U.S. Army Engineer Topographic Laboratories, 1972. vi, 152 p. Available from NTIS, AD 751 192.

This paper is divided into two sections. One identifies the terrain elements to be studied, presents the variables to be interpreted, and then lists the remote sensor applications which may be useful for this particular MGI in a systematic way. Following each listing of terrain elements (hydrologic, vegetation, landform, and cultural elements) is the second section which consists of a bibliography. A matrix evaluation is presented for all sensors and all variables within that major group.

This paper is probably one of the most well-balanced presentations regarding the evaluation of remote sensing presently in the literature. It is written so that any section can be scanned and studied almost independently of any other section of the work. The bibliography is very complete. A few questions arise, but these are rare.

128A Von Bandat, Horst F. AEROGEOLOGY. Houston: Gulf Publishing Co., 1962. xii, 350 p. Figs., Illus., index, refs.

> This loose-leaf publication consists of six major parts dealing with photogrammetry, aero geologic practice, general analysis, morphologic expression of rock types, structural analysis and landform analysis. Numerous stereo pairs are provided together with detailed statements and analysis of each. The text consists of introductory statements which provide the basic concepts and terminology used in the remainder of the separate sections.

129. Watson-Watt, Sir Robert. THREE STEPS TO VICTORY: A PERSONAL ACCOUNT BY RADAR'S GREATEST PIONEER. London: Odhams Press, 1957. 480 p.

> The military backgrounds of radar during and subsequent to World War II are given in this autobiography. This background of radar, although not directly related to present earth resources, gives an appreciation for the amount of work that has been needed to develop the present instruments. Very interesting auxiliary reading.

130. Way, Douglas S. TERRAIN ANALYSIS: A GUIDE TO SITE SELECTION USING AERIAL PHOTOGRAPHIC INTERPRETATION. Community Development Series, R.P. Dober, general editor. Stroudsburg, Pa.: Dowden, Hutchinson and Ross, 1973. viii, 392 p. Illus., tables, figs.

> As the book and series title implies the emphasis in this book is the use of aerial photographs for planning purposes. Nothing is said that deals with photogrammetry, or the philosophies of the methods used; rather the emphasis is on extraction of information from the aerial photographs. Hence, chapters dealing with processes of physical geology and soil types as well as differentiations are the key to the book, as are the latter chapters which discuss various rock types and their appearance on aerial photographs. Much concern is shown for using surrogate information (e.g., drainage patterns) to arrive at the type of information actually desired by site and community planners.

131. Wells, Helen T.; Whiteley, Susan H.; and Karegeannes, Carie E. ORIGINS OF NASA NAMES. The NASA History Series. NASA SP-4402. Washington, D.C.: U.S. National Aeronautics and Space Administration, 1976. x, 227 p. Illus., appendixes, refs., index. Available from the U.S. Government Printing Office.

> This book was designed to answer some questions about the origins of NASA associated names. Some of these names

originated from no specifically identifiable source, others are acronyms, and still others were the result of a very formal process used for naming projects. Six major sections make up the bulk of this publication dealing, in turn, with launch vehicles, satellites, space probes, manned space flights, sounding rockets, and NASA installations. Four appendixes present abbreviations, acronyms, and terms; international designation of spacecraft; NASA major launch records, 1958–74; and NASA naming committees. In this maze of names, acronyms, and terms, this publication is possibly one of the most useful for sorting out the various projects, satellites, and the sequence of their launch, especially since many of the designations do not apparently follow any logical and/or consistent sequence.

132.   Wenderoth, Sondra, and Yost, Edward.   MULTISPECTRAL PHOTOG-
RAPHY FOR EARTH RESOURCES.   Greenvale, N.Y.: Science Engineering Research Group, C.W. Post Center, 1974. 279 p. Tables,
figs., illus., refs., glossary. Available from NTIS, N75-15138.

Gives a good introduction to the physics of light and the nature of films, together with a discussion of the operation of visible remote sensors. Contains numerous examples of imagery and exercises together with some analysis of the various scenes shown. Lacks an index but has extensive table of contents. A guide for producing accurate multispectral results for earth-resource applications is presented along with theoretical and analytical concepts of color and multispectral photography. Topics discussed include the following: (1) capabilities and limitations of color and color infrared films, (2) image color measurements, (3) methods of relating ground phenomena to film density and color measurements, sensitometry, (4) considerations in the selection of multispectral cameras and components, and (5) mission planning.

133.   Westinghouse Electric Co.   SIDE-LOOK RADAR.   Baltimore: September 1967. 46 p. Illus. Paper, (spiralbound).

The high resolution side-looking radar imagery illustrating this brochure was produced by the Westinghouse Electric Company for NASA using the AN/APQ 97 radar developed for the U.S. Army Electronics Command. Certain material on the application of this imagery has been prepared by the Center for Research in Engineering Science, University of Kansas and USGS. This excellent, but difficult to locate, brochure contains over a dozen annotated images illustrating various applications of these data to the geosciences. It also contains brief explanations concerning the operation of such imaging radars.

134. Wheeler, Gershon J. RADAR FUNDAMENTALS. Series in Electronic Technology, edited by I.L. Kosow. Englewood Cliffs, N.J.: Prentice-Hall, 1967. xiii, 105 p. Index, illus., figs.

> This book presents a nontechnical explanation of what radar is and how it works. A chapter is devoted to each major subsystem of the radar system, and different radar systems are explained in terms of these building blocks. The book was specifically written as a text for a one-semester course in radar at the technical institute or junior college level. It should also be useful to the engineer or layman desiring a basic understanding of the subject.

135. White, Leslie P. AERIAL PHOTOGRAPHY AND REMOTE SENSING FOR SOIL SURVEY. Oxford and New York: Clarendon Press, 1977. 118 p. Figs., refs., color and b/w plates.

> This work outlines the use of aerial photography for soil mapping and directs attention to the working principles of cameras and other image-obtaining devices. The ways that the images are produced and the methods of using them in soil surveys are discussed. Both airborne and spaceborne platforms are considered as are several processing techniques (e.g., image enhancement and automatic image analysis). The bulk of the publication deals with the basic concepts of remote sensing systems including imaging radar, multispectral scanners, and thermal imagers. Aerial cameras apparently remain as the primary remote sensor used for soil studies.

135A Widger, William K., Jr. METEOROLOGICAL SATELLITES. New York: Holt, Rinehart and Winston, 1966. 280 p.

> A series of meteorological satellites, starting with TIROS I in 1960 have been continuously operating and returning data to the earth for analysis. This small book describes not only the programs but the rationale and the instruments of the early meteorological satellite systems. It is an excellent publication written in a nontechnical language.

136. Williams, Richard S., Jr., and Carter, William D., eds. ERTS-1: A NEW WINDOW ON OUR PLANET. U.S. Geological Survey Professional Paper 929. Washington, D.C.: U.S. Government Printing Office, 1976. xix, 362 p. Col. plates, illus. Paper.

> Papers prepared primarily by scientists in various U.S. government and state government organizations are presented in eight major chapters. These chapters cover cartography, geology and geophysics, water resources, land use mapping and planning, agriculture, forestry and rangeland, environmental monitoring, conservation, and oceanography. Numerous color plates of ERTS-1 imagery are included.

137. Wolf, Paul R. ELEMENTS OF PHOTOGRAMMETRY: WITH AIR PHOTO INTERPRETATION AND REMOTE SENSING. New York: McGraw-Hill Book Co., 1974. xiii, 562 p. Figs., plates, tables.

   Designed as a text for introductory courses in photogrammetry at the college level, this publication provides a solid introduction to photogrammetry, remote sensing, and photo interpretation with the strongest emphasis on photogrammetry.

138. Wolff, Edward A., and Mercanti, Enrico P., eds. GEOSCIENCE INSTRUMENTATION. New York: John Wiley and Sons, Wiley-Interscience, 1974. xxvi, 819 p. Figs., tables, bibliog., index.

   This publication deals with the basic theory, design, and operation of numerous geophysical instruments, including the array of earth-resources remote sensors which are presently operating from aircraft and satellite platforms. An extremely valuable publication for background on instrumentation. Geoscience instrumentation systems are considered along with questions of geoscience environment, signal processing, data processing, and design problems. Instrument platforms are examined, taking into account ground platforms, airborne platforms, ocean platforms, and space platforms. In situ and laboratory sensors described include those suitable for detecting and recording acoustic waves, atmospheric constituents, biological phenomena, cloud particles, electric fields, electromagnetic fields, precision geodetic phenomena, gravity, ground constituents, magnetic fields, horizons, humidities, ions and electrons, tides, and winds. Remote sensors are discussed, giving attention to sensing techniques, acoustic echo-sounders, and gamma ray, optical, radar, and microwave-radiometric sensors.

139. Zdanovich, V.G., ed. METHODS FOR STUDYING OCEAN CURRENTS BY AERIAL SURVEY. Translated from the Russian by A. Barouch. Jerusalem: Israel Program for Scientific Translations, 1967. vi, 212 p. Figs., illus., tables, refs.

   Based on research done in Russia by the Laboratory of Aerial Methods of the State Geological Committee, this book describes the use of instruments and visual observations for measuring ocean currents. Aerial methods were from aircraft rather than from spacecraft. Particular attention is given to the practical problems encountered in field and laboratory work. Familiarity with the fundamentals of aerial surveying, photography, and photogrammetry are assumed.

Chapter 2
PROCEEDINGS

Many of the papers and the flow of information which are presently experienced by the earth-resources and remote-sensing communities occur in the publication of symposia proceedings, meeting notes, minutes, and similar sources. This is, of course, because many of the developments are rapidly taking place and because of the lack of journals which are devoted solely, or primarily, to remotely sensed data. However, there are many journals available which include remote-sensing papers--these are listed in the journal section, chapter 7, of this information guide. In addition, because many of the symposia are held on a cyclical basis, either annually or semiannually, members of the remote-sensing community have become somewhat accustomed to exchanging their basic ideas and data during these meetings. Hence, the proceedings are the record of such exchanges.

Included in this portion of the bibliography is a listing of the publications which are the outgrowths of such meetings. Many, as indicated in the specific annotations, are recurring, and the reader is advised to seek the most recent sources to determine if these meetings are continuing on a regular basis. Examples of these are the Pecora Meetings (generally held in Sioux Falls, South Dakota); the Purdue conferences dealing with machine processing of remotely sensed data (West Lafayette, Indiana); the University of Michigan conferences titled "Remote Sensing of Environment" (generally held in Ann Arbor, Michigan); and the Arid Land Remote Sensing Conferences (held in Tucson, Arizona).

Like the quality of publications everywhere and in all aspects of the references in this bibliography, the quality of the papers included in this section vary widely. Some symposia organizers simply take the "camera ready" copy provided by the authors and do no editorial work before publication in the proceedings. Consequently, the reader must be cautious when accepting some of the publications at face value. Also, it is noted that one of the major reasons for presenting a paper at a symposia is to allow the authors to express to their peers the nature of their work and to receive guidance, comments, and interaction from their audience with respect to the development, approach, and validity of their scientific efforts. Hence, it is not unusual to find a paper which speaks with a high degree of confidence of the work and then to have the concept almost completely disappear from the following literature. One may assume, in many cases, that the exuberant author was suddenly made aware of the pitfalls of his approach. Hence, we find in the proceedings the papers which often expose the scientists to the criticism of the scientific community and can observe the concept of peer review in action. By the same token, through searching such literature as that which is included in this portion of the bibliography, we are able to trace the initial stages of the development of an idea and to also identify approaches which have been developed but could not be continued because of the need for some technological advance which had not, at that time, been completed.

Proceedings, thus, are excellent sources for the observation of the dynamic nature of any discipline, and in the area of remote sensing, which is so new, these publications are possibly the best single source for maintaining an awareness of the rapid and interesting developments which are occurring.

140. American Institute of Aeronautics and Astronautics. CONFERENCE ON SCIENTIFIC EXPERIMENTS OF SKYLAB. Series of reprints of papers presented at the conference on the scientific experiments of Skylab held 30 October 1974 to 10 November 1974, Huntsville, Ala. New York: 1974. Var. pag. Paper, (spiralbound).

> These papers cover the operational aspects of Skylab, analysis of data collected during the flights, and the comparison of Skylab with other satellite data. Also included are papers dealing with experiments of Skylab which were not oriented toward earth remote sensing.

141. American Society of Photogrammetry. PROCEEDINGS OF THE FALL TECHNICAL MEETING. Falls Church, Va.: 1973-- . Annual. Bound proceedings are available from the ASP office.

> The American Society of Photogrammetry (ASP) sponsors an annual fall meeting each year as part of their program of professional development and interchange of ideas and research. This meeting is held at various locations within the United States during September or October. These proceedings contain fully illustrated (halftone) papers supplied by the various authors (in camera-ready form) to the ASP. Some of the papers are later published in the ASP journal PHOTOGRAMMETRIC ENGINEERING AND REMOTE SENSING.

141A American Society of Photogrammetry. MAPPING WITH REMOTE SENSING DATA. Proceedings, 2d Annual William T. Pecora Symposium. American Society of Photogrammetry and U.S. Geological Survey. Washington, D.C.: American Society Photogrammetry, 1977. vxi, 404 p. Illus., refs., figs., tables. Paper.

> This meeting was held at Sioux Falls, South Dakota, on

25-29 October 1976. According to the proceedings for-word, the papers were to address the question: "How can remote sensing data be referenced, interpreted, and dis-played for effective management decisions?" To this end, thirty papers in six sessions were presented. The sections of the proceedings are as follows: (1) remote-sensing map-ping requirements, (2) data acquisition and processing, (3) mapping of dynamic phenomena, (4) mapping of thematic information, (5) information distribution, and (6) accuracy standards for remote sensing. (See also citation no. 202).

142. _____. TECHNICAL PAPERS FROM THE ANNUAL MEETING. Falls Church, Va.: 1973-- . Annual. Bound proceedings of these meet-ings are available from the ASP office.

The American Society of Photogrammetry sponsors an annual meeting as a regular part of their program of professional development and interchange of ideas and research. The annual meeting is held in Washington, D.C., during Febru-ary or March. These proceedings contain fully illustrated (halftone) papers supplied by the various authors (in camera-ready form) to the ASP. The cost is generally quite rea-sonable. Some of the papers presented are later published in the ASP journal PHOTOGRAMMETRIC ENGINEERING AND REMOTE SENSING.

143. Anson, Abraham, ed. MANAGEMENT AND UTILIZATION OF RE-MOTE SENSING DATA: PROCEEDINGS OF THE SYMPOSIUM. Symposium on Remote Sensing held on 29 October 1973 to 1 November 1973 at Sioux Falls, South Dakota. Falls Church, Va.: American Society of Photogrammetry, 1973. 687 p. Illus., figs., tables.

Applications of remote sensing to the planning of manage-ment of specific earth resources are described along with developments in remote imagery hardware. A total of fifty-two papers were presented at this meeting. Papers are re-produced in the proceedings, and most contain good bibli-ographies and reference lists. Many of the papers present preliminary results of ongoing research.

144. Arctic Institute of North America. SYMPOSIUM ON REMOTE SENS-ING OF THE POLAR REGIONS. Symposium held in Easton, Md., 6-8 March 1968. Washington, D.C.: 1968(?). iii, 66 p. Available from NTIS, N69-22267.

This symposium was held to acquaint scientists and techni-cians concerned with remote sensing with some of the spe-cial problems of the polar areas and, in turn, to acquaint polar scientists with the potential of the use of remote sens-ing. This publication presents some of the highlights of this symposium. The presentations of seven panels were presented. These were: (1) the ionosphere, upper atmosphere, and atmosphere, (2) physical oceanography, (3) biological oceanography, (4) earth resources, geomorphology, glaciology, and permafrost, (5) terrestrial biology, (6) applications and development, and (7) summary and a look ahead. The presentations are from an overview and a conceptual point of view.

145. Arizona, University of. Office of Arid Lands Studies. PROCEEDINGS; CONFERENCE ON REMOTE SENSING IN ARID LANDS. Tucson: 1970-- . Var. pag. Illus., figs., refs. Proceedings of the second, third, and fourth conferences are available from the University of Arizona, Office of Arid Lands Studies, 1201 E. Speedway Blvd., Tucson, Ariz. 85719.

The annual conference on remote sensing in arid lands was initiated in 1970 with the creation of the Arizona Ecological Test Site (ARETS) by the U.S. Department of Interior's Earth Resources Observation System (EROS) program and the University of Arizona. The first conference was the conceptual meeting to establish ARETS and no proceedings were produced. The annual conference is designed to bring together scientists whose remote-sensing applications focus on the arid areas of the world and related problems such as land use, environmental monitoring, agriculture, and forestry as well as range, mineral, and water resources.

146. Barrett, Eric C., and Curtis, Leonard F. ENVIRONMENTAL REMOTE SENSING: APPLICATIONS AND ACHIEVEMENTS. Papers presented at the Bristol Symposium on Remote Sensing, Department of Geography, University of Bristol, 20 October 1972. London: Edward Arnold; New York: Crane, Russak, 1974. vi, 309 p. Illus., figs., refs.

Fifteen papers are presented in three major sections: (1) rock, soil, and landforms, (2) land use, vegetation, and crops, and (3) water, weather, and climate. Papers do not cover the entire field of remote sensing and each paper is essentially independent of all others.

147. Battrick, B.T., and Duc, N.T., eds. EUROPEAN EARTH RESOURCES SATELLITE EXPERIMENTS. Proceedings (in English and French) of a symposium held from 28 January 1974 to 1 February 1974 in Frascata, Italy. Publication ESRO SP-100. Paris: European Space Research Organization, May 1974. 469 p. Illus., bibliog., figs. Available from NTIS, N75-14216.

The major topics were the use of ERTS-1 data and data processing for ERTS, Skylab, and EREP imagery. Applications of these data to the fields of oceanography, hydrology, glaciology, geology, geomorphology, agriculture, forestry, and vegetation are also covered.

148. Blackband, W.T., ed. ADVANCED TECHNIQUES FOR AEROSPACE
SURVEILLANCE. Twenty-three papers presented at the Thirteenth
Symposium of the Avionics Panel of AGARD held in Milan, Italy, on
4-7 September 1967. Conference Proceedings no. 29. Advisory
Group for Aerospace Research and Development (AGARD). Paris:
North Atlantic Treaty Organization, 1968. 425 p. Available from
NTIS, N68-33392.

> Various techniques for surveillance by aircraft and satellites
> are reviewed. Discussed are photography, television, milli-
> meter wavelength side-looking radar, infrared radiometry,
> and millimeter radiometry. Papers are presented in either
> English or French. Papers also concern some interpretation
> techniques and methods of recording and displaying sensed
> data.

149. Bock, Paul, ed. APPROACHES TO EARTH SURVEY PROBLEMS
THROUGH USE OF SPACE TECHNIQUES. Proceedings of the Sym-
posium held in Konstanz, West Germany, 23-25 May 1973. Sympo-
sium sponsored by the Committee on Space Research (COSPAR) and
Deutsche Forschungsgemeinschaft. Berlin, East Germany: Akademie-
Verlag, 1974. 479 p. Illus., figs., refs.

> Reports on the use of remote-sensing techniques for geog-
> raphy, earth surveying, land use mapping, oceanography,
> atmospheric and climatological studies, and meteorology are
> presented. Some of the topics covered include remote sens-
> ing in the study of Antarctic marine resources and verte-
> brates, earth satellite measurements as applied to sea ice
> problems, use of ERTS-1 satellite in remote sensing of wa-
> ter resources in Canada, ERTS imagery for study of snow
> and glacier hydrology, a low-cost system for reproducing
> ERTS imagery, need for and aspects of a cooperative Euro-
> pean earth resources program, and optical investigations
> from the manned spacecraft Voskhod 2. A total of forty-
> two papers is included.

150. California Institute of Technology, Pasadena. Jet Propulsion Laboratory. PROCEEDINGS: CALTECH/JPL CONFERENCE ON IMAGE
PROCESSING TECHNOLOGY DATA SOURCES, AND SOFTWARE FOR
COMMERCIAL AND SCIENTIFIC APPLICATIONS. Conference held
at the California Institute of Technology on 30 November 1976. Pasadena, Calif.: November 1976. vii, 181 p. Paper. Available
from NTIS, N77-14540.

> A total of thirty-seven papers, prepared primarily by JPL
> and CIT personnel and concerned primarily with image pro-
> cessing of both remote-sensing and nonremote-sensing data
> sources, is presented.

151. California, University of Los Angeles. Engineering and Science

Extension. REMOTE SENSING OF ENVIRONMENT. Prepared in co-operation with TRW Systems. Los Angeles: August 1968. 1,034 p. Refs. Available from NTIS, N70-14072.

This volume is one of the earlier sets of notes from a short course. It consists of lectures given during a short study course in remote sensing of the environment with an emphasis on the fundamental concepts, techniques, and equipment, and potential applications. Imaging techniques, specific types of sensors, including side-looking radar systems and multispectral scanners; ground truth requirements; environmental factors affecting system performance; and data processing methods are treated in detail. Fields of application discussed in the various lectures are meteorology, oceanography, earth resources management, geology, and geography.

152.   Canada. Defence Research Board. PROCEEDINGS OF A SEMINAR ON THICKNESS MEASUREMENT OF FLOATING ICE BY REMOTE SENSING. Held on 27-28 October 1970. Technical Note 71-74. Ottawa: Defence Research Establishment, Ottawa. June 1971. x, 260 p. Illus., figs., refs. Paper.

The seminar on thickness measurement of floating ice by remote sensing was organized by the Defence Research Establishment, Ottawa, with the aim of bringing together those working on the difficult problem of remote measurement of ice thickness, and particularly sea ice thickness, and potential users of such methods. It was hoped thus to crystallize some of the problems involved and to stimulate useful discussion between scientists in different fields. It is felt that the seminar succeeded in this aim. Ten papers were presented without editing and as received from the speakers. Several systems (acoustic, pulsed sonic methods, radars, seismic, radiometers, etc.) are discussed and are the subjects of the several papers.

153.   Canadian Aeronautics and Space Institute. REMOTE SENSING OF SOIL MOISTURE AND GROUNDWATER, PROCEEDINGS OF THE WORKSHOP. Workshop held in Toronto, Ontario, Canada, on 8-10 November 1976. Ottawa: 1977. 246 p. Figs., refs., tables. Available from AIAA, A78-18860 to A78-18872, inclusive.

An introduction to hydrologic problems and the principles of remote sensing is presented. Agriculture soil moisture and ground water problems in the prairie provinces of Saskatchewan and Alberta are considered, together with the application of remote sensing to watershed modeling and real-time flood forecasting. Some geophysical techniques which are surface based are also considered along with airborne and spaceborne techniques for study of soil moisture and groundwater content and problems. Landsat data are also considered for some special aspects of this portion of hydrology.

154. Collins, W. Gordon, and Van Genderen, John L., eds. FUNDA-MENTALS OF REMOTE SENSING. Proceedings of the First Technical Session of the Remote Sensing Society, consisting of five formal papers presented at the meeting held on 13 February 1974. London: Remote Sensing Society, February 1974. ii, 149 p. Figs., illus., refs. Paper.

> All papers deal with the basic physical principles of remote sensing for the benefit of specialists in various earth-oriented disciplines. This is a companion volume to Van Genderen and Collins's REMOTE SENSING DATA PROCESSING (chapter 1 as item no. 126).

155. Cooper, Saul, and Ryan, Philip T. DATA COLLECTION SYSTEM: EARTH RESOURCES TECHNOLOGY SATELLITE - 1. Proceedings of a workshop held at NASA Wallops Flight Center, Wallops Island, Va., 30-31 May 1973. NASA SP-364. Washington, D.C.: Scientific and Technical Information Office, National Aeronautics and Space Administration, 1975. vi, 132 p. Paper. Available from NTIS N75-16050.

> Subjects covered at the meeting concerned results of the overall data collection system including sensors, interface hardware, power supplies, environmental enclosures, data transmission, processing and distribution, maintenance and integration in resources management systems. Fourteen technical presentations and seven agency presentations are included in the publication. Provides a good background for the nonimaging aspects of the ERTS (now Landsat) system.

156. Ewing, Gifford C., ed. OCEANOGRAPHY FROM SPACE: PROCEEDINGS. WHOI Ref. no. 65-10. Papers presented at the Conference on the Feasibility of Conducting Oceanographic Explorations from Aircraft, Manned Orbital and Lunar Laboratories held on 24-28 August 1964. Sponsored by the National Aeronautics and Space Administration, contract NsR-22-014-033. Advanced Missions, Manned Space Science Program, Office of Space Science and Applications. Woods Hole, Mass.: Woods Hole Oceanographic Institution, 1965. xxi, 469 p. Illus., figs., charts, maps, bibliog. Available from NTIS, N65-30350.

> These papers consist of a sampling of the opinions of a representative cross section of oceanographers. Areas in physical, biological, chemical, and geographical aspects of oceanography are covered.

157. Ferdman, Saul, ed. THE SECOND FIFTEEN YEARS IN SPACE: PROCEEDINGS OF THE ELEVENTH GODDARD MEMORIAL SYMPOSIUM. Science and Technology Series, vol. 31. Tarzana, Calif.: American Astronautical Society, 1973. 196 p. Illus., refs., figs.

> A series of papers examining the next fifteen years (1973-88)

in space and the potential benefits and impact of the space programs on industrial, scientific, and social aspects of life in the United States is presented. They deal primarily with the technologies and the benefits from developing technologies. Few comments concerning earth resources, per se, are included.

158. Ford, C. Quentin, ed. SPACE TECHNOLOGY AND EARTH PROBLEMS. Science and Technology Series, vol. 23. Proceedings of American Astronautical Society, 23-25 October 1969, symposium at Las Cruces, N.Mex. Tarzana, Calif.: American Astronautical Society, 1969. xiv, 401 p. Illus.

This symposium was "devoted to showing that solutions developed for space problems have direct application to earth problems. . . ." These applications included atmospheric, communications, land, ocean and water and transportation systems, and sensors and data management. Of special importance for earth-resources remote sensing is a series of eleven papers dealing with spaceborne remote sensors. Four of these are presented in their entirety.

159. Freden, Stanley C.; Mercanti, Enrico P.; and Becker, Margaret A. SYMPOSIUM ON SIGNIFICANT RESULTS OBTAINED FROM THE EARTH RESOURCES TECHNOLOGY SATELLITE - 1. 3 vols. NASA SP-327. Proceedings of a symposium held by the Goddard Space Flight Center at New Carrollton, Md., 5-9 March 1973. Washington, D.C.: U.S. Government Printing Office, May 1973. Paper. Available from NTIS, N73-28207, N73-28389, and N73-28405.

This symposium provided the first open forum where the users of the ERTS data had an opportunity to present the significant accomplishments of their investigations. It also provided the first opportunity for representatives of federal, state, and local organizations to present their views on how ERTS data are being used and will be used for solving operational resource management problems. A total of 156 papers is presented. Contents are as follows: Volume 1, TECHNICAL PRESENTATIONS, volume 2, SUMMARY OF RESULTS, and volume 3, DISCIPLINE SUMMARY REPORTS.

160. Freden, Stanley C., and Friedman, David B., eds. THIRD EARTH RESOURCES TECHNOLOGY SATELLITE SYMPOSIUM. 5 vols. Third Symposium on Significant Results Obtained from the First Earth Resources Technology Satellite was held 10 December 1973 to 14 December 1973 in Washington, D.C. NASA SP-351, SP-356, SP-370. Washington, D.C.: U.S. Government Printing Office, 1974. Illus., figs., refs. Paper. Available from NTIS.

Volume 1 contains the proceedings of the opening plenary sessions, volume 2 contains the summaries of the results of research, and volume 3 consists of the discipline summary reports. Papers cover the areas of agriculture and forestry as well as range resources, mineral resources, geological structure, and landform surveys, water resources, marine resources, environment surveys, and interpretation techniques. The volumes are arranged thus:

Vol.				
1.	TECHNICAL PRESENTATIONS - SECTION A	975 p.	N74-30705
	TECHNICAL PRESENTATIONS - SECTION B	936 p.	N74-30774
2.	SUMMARY OF RESULTS	179 p.	N74-10549
3.	DISCIPLINE SUMMARY RESULTS	155 p.	N74-33873

An additional volume containing the abstracts only (NTIS N74-11161) is also available.

160A   Greve, T.; Lied, F.; and Tandberg, E., eds. THE IMPACT OF SPACE SCIENCE ON MANKIND. New York: Plenum Press, 1976. vii, 125 p. Figs.

The proceedings of the thirty-first Nobel Symposium held at Spatind, Norway, September 7-12, 1975, are published in this volume as a set of eight edited summaries of papers presented. Summaries of the discussions are also included. Four major areas were discussed, being the impact of space science, the impact of space communications, the impact of earth resources exploration from space and the impact of space assisted meteorology. Although not directly dealing with remote-sensing imagery and data, the papers and discussions give very significant insights into the ways satellite techniques (including remote-sensing imagers) are altering our views of the world and aiding in our understanding of the earth.

161.   Institute of Electrical and Electronics Engineers. INTERNATIONAL RADAR CONFERENCE. Conference held from 21-23 April 1975 in Arlington, VA. New York: 1975. 639 p. Illus., figs., refs.

This conference considered the proposed developments in radar technology from 1975 to 1985. Although many of the papers presented deal with nonimaging radars and with techniques which are not directly applicable to earth resources and earth science studies, the outlines of radar development during the studied time-frame do indicate some of the new instruments which shall become available to earth scientists.

162.   Institute de Pesquisas Espaciais. SEMINAR ON SPACE APPLICATIONS OF DIRECT INTEREST TO DEVELOPING COUNTRIES. 2 vols. Conference held in Brazil on 16-19 June 1974. São José dos Campos,

Brazil: 1974. 372 p. Illus., figs., refs.

This conference concerned itself with the application of
remote-sensing techniques in the surveying of earth resources
in developing countries. These techniques are described as
applied to specific problems and are also presented from the
viewpoint of their effectiveness as a basis for socioeconomic
development. Some of the topics covered include the application of ERTS results in the Republic of South Africa,
acquisition and use of ERTS-1 data by resources management
in Brazil, hydrogeologic evaluation of ERTS and EREP data
for the pampa of Argentina, human settlement patterns in
relation to resources of lesser developed countries, and tec-
tonic interpretation of ERTS-1 mosaics of the La Paz area
of Bolivia.

163. Instrument Society of America. SECOND JOINT CONFERENCE ON
SENSING OF ENVIRONMENTAL POLLUTANTS. Proceedings of a
meeting held on 10-12 December 1973. Pittsburgh, Pa.: 1973. iv,
377 p. Figs., illus., refs.

This meeting, cosponsored by a group of nine professional
and government organizations, was the source of these proceedings. These cosponsors were the American Chemical
Society, American Institute of Aeronautics and Astronautics,
American Meteorological Society, U.S. Department of Transportation, Environmental Protection Agency, IEEE, Instrument
Society of America, NASA, and NOAA. A total of forty-
six papers were presented in thirteen major sessions. Although several papers dealt with in situ measurements, many
were concerned with atmospheric and water pollution on the
local, regional, and global scales. Some detailed examples
of remote-sensing applications using very specific instrumentation are included.

164. Interdepartmental Committee on Air Surveys. PROCEEDINGS OF THE
SECOND SEMINAR ON AIR PHOTO INTERPRETATION IN THE DE-
VELOPMENT OF CANADA. Symposium held on 13-15 March 1967
in Ottawa, sponsored by the Interdepartmental Committee on Air Surveying. Ottawa: Queen's Printer, 1968. iv, 213 p. Illus., figs.,
bibliog. Paper.

The accent of this symposium was on photo interpretation as
it relates to the utilization of land resources and on the
education for present and future photo interpreters. In in-
viting the papers for this symposium, an attempt was made
to illustrate problems of technical methodology and the lack
of trained staff, and, where possible, to have the papers
indicate the probable direction in which the answers will be
found. A total of twenty papers are included in the publication.

165. International Glaciological Society. A Symposium on Remote Sensing in Glaciology, Cambridge, England, 16-20 September 1974. JOURNAL OF GLACIOLOGY 15, no. 73 (1975): 1-482. Paper.

> Papers are presented describing applications of remote-sensing techniques including radar, laser, thermal infrared, multispectral scanning and sonar, in the study of glaciers, sea ice, pack ice, and other glaciological phenomena. Some of the topics covered include theory of radio echoes from glacier beds, ultrasonic properties of plastically deformed ice, comparison of sea ice type identification between airborne dual frequency passive microwave radiometry and standard laser and infrared techniques, analysis of the near-surface energy transfer environment from thermal infrared imagery, the optical properties of salt ice, and the use of ERTS images to measure the movement and deformation of sea ice. A total of thirty complete papers are published, in addition to ten abstracts of papers presented but not published in full, and ten additional abstracts of papers accepted for presentation but not read. Exceptionally good bibliographies are included with the papers. (See citation no 347.)

166. International Society for Photogrammetry. Commission VII. PROCEEDINGS: SYMPOSIUM ON REMOTE SENSING AND PHOTO INTERPRETATION. 2 vols. Proceedings of a meeting held in Banff, Alberta, 7-11 October 1974. Ottawa: Canadian Institute of Surveying, 1974. 878 p. Figs., illus., tables, refs.

> The technical program of the symposium concentrated on the application of photo interpretation and remote sensing to resource inventories and environmental studies, including the following discipline areas: land use, resource inventories, water and wetlands monitoring, environmental monitoring, vegetation damage, and geology. Attention is paid to the interpretation and analysis of remote sensor images. Landsat multispectral scanner observations are applied to mapping urban land use in the United States, to soil mapping in California, and to forest classification at the regional level. Aerial photography is applied to terrain mapping, to the assessment of volume characteristics of tropical rain forests and to forest monitoring. A total of sixty-eight papers in French, English, and German (English predominates), are included in the two volumes.

167. Iowa. Geological Survey. PROCEEDINGS, SEMINAR IN APPLIED REMOTE SENSING. Public Information Circular no. 3. Fifteen papers presented at the Applied Remote Sensing seminar held at Drake University on 8-12 May 1972 in Des Moines. Iowa City: September 1972. ii, 181 p. Figs, illus. Paper.

> The papers deal with the basic principles of various remote-

sensing instruments and the application of some remote-sensing data to problems encountered in the state of Iowa.

168. Johnson, Philip L., ed. REMOTE SENSING IN ECOLOGY. Proceedings of a symposium held 19 June 1968, at Madison, Wis., sponsored by the Ecological Society of America and the American Society of Limnology and Oceanography. Athens: University of Georgia Press, 1969. x, 244 p. Illus., maps, bibliog.

A total of fourteen contributed papers covering electromagnetic remote sensing and dealing with applications rather than instrumentation and hardware are included in this publication. The stress is on the plant communities although animal communities and their study by remote sensing are included.

169. Joint Publications Research Service. MEETING OF THE SOVIET-AMERICAN WORKING GROUP ON REMOTE SENSING OF THE NATURAL ENVIRONMENT FROM SPACE   [Vstrecha Sovetsko-Amerikanskoy Rabochey Groppy po Issledovaniyu Okuzhayushchey Sredy iz Kosmosa]. Papers from the meeting of the Soviet-American Working Group on Remote Sensing of the Natural Environment from Space which was held in Moscow on 12-17 February 1973. JPRS - 58739. Moscow: Academy of Science, Arlington, Va.: 1973. 193 p. Figs., refs. Available from NTIS, N73-30317.

This report consists of the translation of a series of Russian-language papers dealing with the geological interpretation of images of the earth taken from outer space and their application for mapping the natural environment.

170. Katz, Y.H., ed. OPTICAL INSTRUMENTATION IN SCIENCE, TECHNOLOGY AND SOCIETY. Proceedings of the Sixteenth Annual Technical Meeting of Society of Photo-Optical Instrumentation Engineers, 1-18 October 1972, vol. 4. Redondo Beach, Calif.: Society of Photo-Optical Instrumentation Engineers, 1972. 282 p.

Generally of limited utility to the earth resources community, this publication presents papers and discussions of mini-computer image storage, data handling instrumentation, sensors, and pattern recognition. Primarily very technical in nature.

171. Lawrence, Mary Margaret, ed. CENTO SEMINAR ON THE APPLICATIONS OF REMOTE SENSORS IN THE DETERMINATION OF NATURAL RESOURCES. Seminar held at Ankara, Turkey, on 10-13 November 1971. Ankara, Turkey: Public Relations Division, Central Treaty Organization, 1971. 179 p. Illus., figs., refs. Paper.

A total of sixteen scientific papers and several other addresses are presented. The objectives of the seminar were to provide the broadest possible base of understanding for systems and knowledge that might be most immediately pertinent to the needs and interests of countries other than the United States.

172. Layton, J. Preston, ed. PROCEEDINGS OF THE PRINCETON UNIVERSITY CONFERENCE ON AEROSPACE METHODS FOR REVEALING AND EVALUATING EARTH'S RESOURCES. Ninety-seventh meeting of the Princeton University Conference, 25-26 September 1969. Princeton, N.J.: Princeton University Conference, June 1970. Var. pag. Figs., refs. Paper.

The proceedings consist of papers delivered at the conference which concern the use of aircraft and spacecraft for sensing earth resources. The general topics covered are advanced remote sensor technology, aircraft and spacecraft systems, data management, data requirements from the user's standpoint, and economic and international aspects.

173. Lomax, John B., ed. PROPAGATION LIMITATIONS IN REMOTE SENSING. Conference Proceedings no. 90. Seventeenth Symposium of the Electromagnetic Wave Propagation Panel of AGARD held at Colorado Springs, Colorado, 21-25 June 1971. Paris: Advisory Group for Aerospace Research and Development (AGARD), 1971. 424 p. Available from NTIS, N72-16085.

Theoretical and experimental performance analyses are reported for various remote-sensing systems in order to develop their propagation ranges and suitabilities in relation to investigated media. Results cover the spectrum from optical to radio frequencies.

174. Matte, Nicolas Mateesco, and DeSaussure, Hamilton. LEGAL IMPLICATIONS OF REMOTE SENSING FROM OUTER SPACE. Symposium sponsored by the Institute of Air and Space Law, McGill University, Montreal, Que., Canada, on 16-17 October 1975. Leyden, Netherlands: A.W. Sijthoff, 1976. xiv, 197 p.

The major function was to analyze the activity and the international legal implications of remote-sensing technology. A total of twenty-one papers are included in five sections: (1) Technical Applications of Remote Sensing from Outer Space, (2) Impact of Remote Sensing on the Economic Development of Western Europe and Latin America, (3) Worldwide Utilization and Dissemination of Data Acquired through Remote Sensing, (4) Possible Integrated North American Landsat Program, and (5) Role of the United Nations.

175. Matthews, Richard E., ed. ACTIVE MICROWAVE WORKSHOP REPORT. NASA SP-376. Washington, D.C.: U.S. Government Printing Office, 1975. xii, 501 p. Available from NTIS N76-11811.

Data from a conference on active microwave systems are summarized. Summaries cover remote sensing of earth and land features, ocean and atmosphere interactions as well as equipment and instrument technology. This volume is somewhat difficult to read due to the lack of any comprehensive index and because it was written by a number of individuals (whose identities are not noted). Hence, when the various portions were blended together, some variance will be noted in notation, style, and so forth. Nonetheless, this report is one of the more up-to-date comprehensive statements on the state-of-the-art vis-a-vis imaging radar remote sensing as applied to earth resources.

176. Michigan, University of. PROCEEDINGS OF THE INTERNATIONAL SYMPOSIUM ON REMOTE SENSING OF ENVIRONMENT. Ann Arbor: University of Michigan, Willow Run Laboratories, 1962-72; University of Michigan and the Environmental Research Institute of Michigan, 1973-- . Illus., figs., charts, refs. Paper.

Held on the average every eighteen months, this is one of the best-known and longest-running remote-sensing symposia. Excellent papers covering all aspects of remote sensing (applications, instrumentation, concepts, legality and economics, among others). Several issues are available from NTIS on microfiche as indicated below. The meetings were first called "Symposium on Remote Sensing of Environment." Starting with the eleventh symposia, the publisher is listed as ERIM.

SYMPOSIA	DATES HELD	PAGES	NTIS NUMBER	PRINTING DATES
First	13-15FEB62	x,110 p.		2d April 1964
Second	15-17OCT62	xiv,459 p.		February 1963
Third	14-16OCT64	xiii,821 p.	N65-33550	2d November 1965
Fourth	12-14APR66	xl,871 p.	N67-13461	June 1966
Fifth	16-18APR68	xlviii,946 p.	N69-33626	September 1968
Sixth	13-16OCT70	xlix,1349 p. (2 vols.)		n.d.
Seventh	17-21MAY71	liii,2366 p.		n.d.
Eighth	2-6OCT72	xlix,752 p. (2 vols.)	N74-18060 or N73-30370	October 1972
Ninth	15-19APR74	xlix,2139 p. (3 vols.)		n.d.
Tenth	6-10OCT75	xlix,1456 p.		n.d.
Eleventh	25-29APR77	xlix,1668 p. (2 vols.)	N78-14529, N78-14464	1977
Twelfth	20-26APR78	lxvi,2,386 p. (3 vols.)		1978

177. Michigan, University of. Institute of Science and Technology. Willow Run Laboratories. THE UNIVERSITY OF MICHIGAN NOTES FOR A PROGRAM OF STUDY IN REMOTE SENSING OF EARTH RESOURCES.

Report no. 1672-1-X. Ann Arbor, Mich.: November 1968. Var. pag. Figs., illus., refs. Paper.

These notes, prepared for a meeting conducted by the University of Michigan at the National Aeronautics and Space Administration Manned Spacecraft Center on 14 February–3 May 1968, were edited, revised, and presented as this report. These notes are essentially the lecture notes; the lectures were divided into several sessions: Fundamentals and Physical Principles; Radiometry and Spectrometry; Photography, Television and Photogrammetry; Scanner, Target Signatures, and Ground Measurements; Radar, Lasers and Passive Microwaves; and Data Processing. This is an excellent set of notes; unfortunately, little imagery is included with the notes.

178. Morgenthaler, George W., and Bursnall, W.J., eds. SPACE SHUTTLE PAYLOADS: PROCEEDINGS OF THE SYMPOSIUM. Science and Technology Series, vol. 30. Tarzana, Calif.: American Astronautical Society, 1973. 509 p. Figs., illus., refs.

This is a series of papers discussing the characteristics of the Space Shuttle system, its payloads, and utilization. Information concerning various plans for scientific projects to be conducted from the shuttle and cost-effectiveness studies of this instrument are included.

179. Morgenthaler, George W., and Morra, Robert, eds. PLANNING CHALLENGES OF THE 70'S IN SPACE. Advances in the Astronautical Sciences, vol. 26. Proceedings of the American Astronautical Society 15th Annual and Operations Research Society 35th Annual meetings, 17-20 June 1969, at Denver, Colorado. Tarzana, Calif.: American Astronautical Society, 1970. xxiv, 445 p.

Five papers on earth orbiter resources missions, in the areas of earth resources, forestry and range resources, agriculture, oceanography, and information services are presented.

179A Plevin, J.; Hood, V.; and Guyenne, T.D., eds. EARTH OBSERVATION FROM SPACE AND MANAGEMENT OF PLANETARY RESOURCES. Workshop held at Toulouse, 6-11 March 1978, sponsored by the Council of Europe, the Commission of the European Communities and the European Association of Remote Sensing Laboratories. Special Report--134. Paris: European Space Agency, May 1978. x, 661 p. Figs., illus., tables, refs.

This document contains the contributions presented at the International Conference on Earth Observation from Space and Management of Planetary Resources. Numerous results of research in the fields of geology, hydrology, irrigation, cartography, as well as current and future European, U.S., Canadian and Soviet programs are presented. During specialized sessions, a survey was made of progress both in the understanding of physical processes of remote sensing and in

the field of algorithmic research. The technical sessions
(dealing mainly with digital processing, microwave, visible
and infrared techniques) showed that high resolution can be
achieved in the near future while keeping a synoptical
view, thanks to the quality of optics and new types of sensors. A session devoted to political and legal implications
followed by a round table discussion on the economical impact of space remote sensing conclude the conference proceedings.

180. PROCEEDINGS IN PRINT. Post Office Box 247, Mattapan, Mass.:
02126. 1966-- . Six times per year. ISSN 0032-9568.

This is an index to conference proceedings in all subject
areas and in all languages. Included are reports of proceedings of conferences, symposia, lecture series, con-
gresses, hearings, seminars, courses, institutes, colloquia,
meetings, and published symposia. Each conference appears
under its unique title, and, wherever possible, place, date,
and sponsorship of the conference, together with ordering
information, are included. Remote sensing is one of the
keywords in the index. Back issues are available from the
publisher.

181. Purdue University. Laboratory for Applications of Remote Sensing.
SYMPOSIUM ON MACHINE PROCESSING OF REMOTELY SENSED
DATA. West Lafayette, Ind.: 1973-- . Var. pag. Illus., figs.,
refs.

This is a series of annual meetings dealing with the processing, management analysis, and applications of remote-
sensing data as applied to the earth sciences. Geographic
information systems, image processing, mathematical pro-
gramming techniques, and pattern recognition techniques are
generally considered in a series of papers presented by individual researchers. These proceedings are generally published by the IEEE and are available from their offices in
New York. This is an excellent annual meeting and the
proceedings should be studied by all remote-sensing scientists interested in computer and machine analysis techniques
and applications.

182. Rango, Albert, ed. OPERATIONAL APPLICATIONS OF SATELLITE
SNOWCOVER OBSERVATIONS. Workshop held on 18-20 August 1975
at the Waystation, South Lake Tahoe, California, sponsored by the
National Aeronautics and Space Administration and the University of
Nevada at Reno. NASA SP-391. Washington, D.C.: National
Aeronautics and Space Administration, 1975. viii, 430 p. Figs.,
illus., refs. Paper.

A total of twenty-eight scientific papers were presented.

They cover photo interpretation and digital techniques used in analyzing Landsat and NOAA satellite data, the use of snow cover observations in stream-flow forecasting and new advances in extraction of snow-pack parameters other than snow extent, employing a variety of remote sensors.

183. Schanda, Erwin, ed. SPECIALISTS MEETING ON MICROWAVE SCATTERING AND EMISSION FROM THE EARTH. Proceedings of the URSI, Commission II, meetings held in Bern, Switzerland, 23-26 September, 1974. Bern, Switzerland: Institute of Applied Physics, University of Bern, 1974. 329 p. Figs., refs., plates. Paper.

The papers address the use of satellites for making surveys of the surface and underground features of the earth, particularly with regard to microwave emission from water, sea ice, land ice, snow, soil, vegetation, and certain geological features.

184. Schneider, William C., and Hanes, Thomas E., ed. THE SKYLAB RESULTS. 2 vols. Advances in the Astronautical Sciences, vol. 31, pt. 1 and 2. Proceedings of the American Astronautical Society 20th Annual Meeting, 20-22 August 1974, at Los Angeles, Calif. Tarzana, Calif.: American Astronautical Society, 1975. Illus., figs., refs. Available on microfiche from Univelt, San Diego, Calif.

This publication contains six papers, all in part 2, concerning the Earth Resources Experiment Package aboard Skylab.

185. Shahrokhi, Firous, ed. REMOTE SENSING OF EARTH RESOURCES. Tullahoma, Tenn.: University of Tennessee Space Institute, 1972-- . Var. pag. Figs., illus., refs. Various volumes are available from Business Office, University of Tennessee Space Institute, Tullahoma, Tenn. 37388.

These volumes are the proceedings of the annual conference on earth resources observation and information analysis system held in Tullahoma, Tennessee. The papers presented here cover the entire field of electromagnetic remote sensing and present preliminary conclusions and statement of ongoing research together with final conclusions of completed work.

186. Tarnocai, C., ed. EVALUATION OF 1971 REMOTE SENSING ACTIVITY IN MANITOBA. Six papers presented at the Fifteenth Manitoba Soil Science Meeting on 8-9 December 1971 held in Winnipeg. Winnipeg: Manitoba Ad Hoc Committee, Resource Satellites and Remote Airborne Sensing, December 1972. 63 p. Figs., tables. Available from NTIS, N72-31382.

This short report consists of some details of the CF-100 aircraft program (consisting of multispectral photography) and

the light aircraft program (Thermal IR scanning) are discussed. An interesting source for discussion of a set of remote-sensing experiments on a provincial scale.

187. Texas A & M University. PROCEEDINGS OF THE SYMPOSIUM ON
REMOTE SENSING IN MARINE BIOLOGY AND FISHERY RESOURCES.
Symposium held on 25 and 26 January 1971, sponsored by Remote Sen-
sing Center and Sea Grant Program Office. College Station, Tex.:
Remote Sensing Center, Texas A & M University, March 1971. 305 p.
Illus., figs., refs. Available from NTIS, N72-30319.

The objectives of this conference were two-fold: (1) to
bring together the investigators active in the utilization of
remote sensing in marine biology and fisheries and (2) to
provide for discussions leading to improved harvest and management of these resources. Fifteen papers are included in
the proceedings.

188. Thompson, G.E., ed. THIRD CANADIAN SYMPOSIUM ON REMOTE
SENSING. Third Canadian Symposium on Remote Sensing held on 22-
24 September 1975 in Edmonton, Alberta, Canada. Ottawa: Canadian
Aeronautics and Space Institute, [1976?]. iv, 516 p. Figs., tables,
refs.

A total of forty-five papers are included, in either English
or French. Agricultural, forestry, and arctic topics domi-
nate the subjects covered with aerial photography and Landsat quite prominent as the sensors used. Other sensors and
subjects are, however, included in the papers and presentations.

189. Thomson, Keith P.B.; Lane, Robert K.; and Csallany, Sandor C., eds.
REMOTE SENSING AND WATER RESOURCES MANAGEMENT. Amer-
ican Water Resources Association Proceedings Series no. 17. Proceedings of a conference held at the Canada Centre for Inland Waters,
Burlington, Ontario, Canada, from 11-14 June 1973. Urbana, Ill.:
American Water Resources Association, 1973. vii, 437 p. Illus.,
figs., maps, refs.

This book is the result of this conference which was attended
by over two hundred scientists and other individuals in the
field of water resources management and remote sensing.
Questions concerning water resources management in Canada
are considered together with the evaluation of environment
quality, the mapping of the 1973 Mississippi river floods
from ERTS, recent applications of remote sensing to water
resources in Hawaii, and the use of ERTS-1 imagery in the
national program for the inspection of dams. Other topics
include the analysis of the drainage pattern of selected areas
of Canada using ERTS-1 imagery as a base, and four wave-
length light detection and imaging systems applied to the

determination of chlorophyll, a concentration and algae
color group, and the use of remote sensing for the mapping
of aquatic vegetation in the Kawartha lakes. Digital processing techniques in thermal plume analysis are examined
along with the remote sensing of stream flow rates.

190.   Tomlinson, R.F., ed. PROCEEDINGS OF THE COMMISSION ON
GEOGRAPHICAL DATA SENSING AND PROCESSING, MOSCOW,
1976. Meeting held by the Commission on Geographical Data Sensing
and Processing of the International Geographical Union was held during
the 23d International Geographical Congress in Moscow, 1976. n.p.:
International Geographical Union, 1977. 136 p. Figs., tables, refs.

The objective of this meeting was to review recent developments in the gathering and handling of geographic data,
particularly the development of geographic information systems and their use as a basis for regional and national planning. A total of seventeen papers, all presented in this
publication in English, is included. These papers address
the collection, processing and interpretation of the geographic
data. Nine of the papers in these proceedings were contributed by Soviet authors.

191.   UNESCO. AERIAL SURVEYS AND INTEGRATED STUDIES, PROCEED-
INGS OF THE TOULOUSE CONFERENCE. EXPLORATION AERIENNE
ET ETUDES INTEGREES, ACTES DE LA CONFERENCE DE TOULOUSE.
Natural Resources Research, vol. 6. Conference on Principles and
Methods of Integrating Aerial Survey Studies of Natural Resources for
Potential Development, 21-28 September 1964, held at Toulouse,
France. Organized by UNESCO, C.N.R.S., and the University of
Toulouse. Paris: 1968. 575 p. Figs., illus., tables. Available
from UNESCO, no. SC/AVS.66/XII.6/AF.

One hundred ninety scientists from forty-five countries met
to make an overall appraisal of the experience gained in
the past twenty years in the use of aerial photography for
the study of natural resources. Their papers, presented in
both English and French, dealt with methods of ensuring the
best use of aerial photography and with the application of
those methods to integrated surveys.

192.   United Nations. SPACE EXPLORATION AND APPLICATIONS. 2 vols.
United Nations Conference on the Exploration and Peaceful Uses of
Outer Space was held at the Kongress-Zentrum in the Hofburg Palace,
Vienna, from 14-27 August 1968. New York: 1969. 1288 p. U.N.
Sales nos. E.F.R.S.69.I.16 Vol. 1 and 2. Paper.

The objectives of the conference were the following: (1)
to examine the practical benefits of space research and exploration on the basis of scientific and technical achievements, and the extent to which nonspace powers, especially

developing countries, might enjoy these benefits, and (2) to examine the opportunities available to nonspace powers for international cooperation in space activities, taking into account the extent to which the United Nations might play a role. Governments, specialized agencies, and other organizations submitted a total of 188 papers. Four languages (English, French, Spanish, and Russian) were used, with the paper included in the original language with summaries in the other three. The papers are grouped in several sections: introduction, communications, meteorology, navigation, other space techniques of practical benefit, and biology and medicine. Because it was an early conference (1968), many of the presently accepted techniques and benefits were not discussed--still, the papers present a good background for helping to understand the origins of the lines along which space exploration and remote sensing of earth-resources have developed.

193. U.S. Army Engineering Topographic Laboratories. PROCEEDINGS OF WORKSHOP FOR ENVIRONMENTAL APPLICATIONS OF MULTISPEC-TRAL IMAGERY. Held at Ft. Belvoir, Va., on 11 November 1975 to 13 November 1975, sponsored by the U.S. Army Topographic Laboratories and the American Society of Photogrammetry. Ft. Belvoir, Va.: 1975. 313 p.

The objectives of the workshop were to identify specific applications of multisensor, multiband, and multispectral imagery, and imagery processing methods to remote-sensing problems associated with land use classification, water and air pollution, exploitation of natural resources, and military geographic information. Toward this end, a total of twenty-one papers were presented at the workshop. All papers dealt with visible and near visible wavelengths and are included in the publication.

194. U.S. National Aeronautics and Space Administration. INTERNA-TIONAL WORKSHOP ON EARTH RESOURCES SURVEY SYSTEMS. 2 vols. Proceedings of a workshop sponsored by the National Aeronautics and Space Administration, Department of Agriculture, Department of Commerce, Department of Interior, Department of State, Agency for International Development and the Department of the Navy, held on 3-14 May 1971 in Ann Arbor, Michigan. NASA SP-283. Washington, D.C.: Government Printing Office, 1971-72. Refs., illus., figs. Paper. Available from NTIS, N73-16348 and N72-16382, and as a microcard from Readex Microprint.

The subjects discussed are (1) aerospace methods of revealing and evaluating the earth resources, (2) remote sensing in geology, hydrology, and geography, (3) applications of remote sensing in agriculture and forestry, (4) remote sensing in oceanography, and (5) use of remote sensing for the study of environmental quality.

195. U.S. National Aeronautics and Space Administration. Goddard Space Flight Center. ADVANCED SCANNERS AND IMAGING SYSTEMS FOR EARTH OBSERVATIONS. Report of a working group meeting at Cocoa Beach, Florida, 11-15 December 1972. NASA SP-335. Washington, D.C.: U.S. Government Printing Office, 1973. xvii, 604 p. Illus., figs., charts, refs. Available from NTIS, N74-11287.

The working group consisted of five panels: (1) electromechanical scanners, (2) self-scanned solid state sensors, (3) electron-beam images, (4) sensor related technology, and (5) use applications. It is concerned primarily with the engineering aspects of remote sensing sensors. It contains a listing of acronyms (most helpful), a short index, and several reprinted papers. Assessments of present and future sensors and sensor related technology are reported along with a description of user needs and applications. Recommendations, charts, system designs, technical approaches, and bibliographies are included for each of the five areas. A total of 122 references is included.

196. U.S. National Aeronautics and Space Administration. Johnson Space Center. PROCEEDINGS: NASA EARTH RESOURCES SURVEY SYMPOSIUM. 3 vols. Proceedings and summaries of the earth resources survey symposium, sponsored by the NASA Headquarters Office of Applications and held in Houston, Texas, 9 June 1975 to 12 June 1975. Houston: 1975. Illus., figs., refs. Paper. Available from NTIS.

Topics include the use of remote-sensing techniques in agriculture, geology, environmental monitoring, land use planning, and management of water resources and coastal zones. Details are provided about services available to various users. Significant applications, conclusions, and future needs are also discussed. The several volumes are as follows:

Vol. 1-A TECHNICAL SESSION PRESENTATIONS
AGRICULTURE - ENVIRONMENT
xvi, p. 1-598   N76-17469

Vol. 1-B TECHNICAL SESSION PRESENTATIONS
GEOLOGY - INFORMATION SYSTEMS
AND SERVICE
xvi, p. 599-1479   N76-17501

Vol. 1-C TECHNICAL SESSION PRESENTATIONS
LAND USE - MARINE RESOURCES
xvi, p. 1498-2166   N76-17552

Vol. 1-D TECHNICAL SESSION PRESENTATIONS
WATER
xvi, p. 2167-685   N76-17588

Vol. 2-A   SPECIAL SESSION PRESENTATIONS
            PLENARY SUMMARIES
            163 p.   N76-26631

Vol. 2-B   SPECIAL SESSION PRESENTATIONS
            COASTAL ZONE MANAGEMENT, STATE
            AND LOCAL USERS, USERS SERVICES
            222 p.   N76-26646

Vol. 3   ABSTRACTS
            xv, 315, p.   N76-17613

197.   U.S. National Aeronautics and Space Administration. Scientific and
        Technical Information Division. REMOTE SENSING OF THE CHESA-
        PEAKE BAY. Proceedings of the Conference on Remote Sensing of
        the Chesapeake Bay held at Wallops Station, Va., 5-7 April 1971.
        NASA SP-294. Washington, D.C.: U.S. Government Printing Office,
        1972. viii, 178 p. Available from NTIS, N72-26272 and N72-
        26285, and as a microcard from Readex Microprint.

        The objective of this conference was to identify the primary
        environmental problems of the Chesapeake Bay area and de-
        termine the extent to which remote sensing can contribute
        to the solution of these problems. The major problem areas
        are environmental pollution, environmental balance, natural
        resources, and other economic activities. A total of twenty-
        three papers, including summaries of group discussions, is
        included.

198.   _____. SIGNIFICANT ACCOMPLISHMENTS IN SCIENCE. Prepared
        by the Goddard Space Flight Center at Greenbelt, Md. Washington,
        D.C.: 1968-- . Figs., illus.

        This series of publications consists of an almost verbatim tran-
        script of the papers presented at the annual symposia. Many
        aspects of NASA work (e.g., oceans, galactic structure,
        astronomy, cometary physics, solar, and terrestrial physics)
        including earth-resources remote sensing are the subjects of
        the papers. Although many of the papers dealing with earth
        remote sensing are specific and somewhat restricted in their
        identified applications, the scope of the papers presented is
        easily extrapolated to other (similar) applications in related
        disciplines. All the volumes are titled as shown above and
        are published as part of the NASA SP series. Some of those
        volumes presently available include:

YEAR	DATE OF SYMPOSIA	NASA SP NUMBER	NTIS NUMBER	PAGINATION
1968	10 Jan. 1969	SP-195	N69-38951	viii, 188 p.

1969	3-4 Dec. 1969	SP-251	N71-25256	xi, 362 p.
1970	15 Jan. 1971	SP-286	N72-23324	vi, 247 p.
1971	10 Nov. 1971	SP-312	N73-13829	viii, 213 p.
1972	7-8 Nov. 1972	SP-331	N73-31867	ix, 233 p.
1973	18-19 Dec. 1973	SP-361	N75-16422	xiii, 348 p.
1974	5 Dec. 1974	SP-384	N76-10934	ix, 200 p.

199.   Veziroglu, T. Nejat, ed. REMOTE SENSING: ENERGY-RELATED
STUDIES. Advances in Thermal Engineering, 5. Proceedings of the
Symposium on Remote Sensing Applied to Energy Related Problems held
on 2-4 December 1974 at the University of Miami in Florida, sponsored
by the Clean Energy Research Institute and the School of Continuing
Studies, University of Miami, Coral Gables, Florida. Washington,
D.C.: Hemisphere Publishing Co., 1975. x, 491 p. Bibliog., index.

   Twenty-two papers in five sections (atmospheric and hydro-
   spheric measurements, active sensor applications land use
   monitoring, environmental quality monitoring, and special
   topics) are included in this publication. All remote-sensing
   systems operating in the EM spectrum are considered and
   numerous examples of their applications and use are included
   in the proceedings.

200.   Von Puttkamer, Jesco, and McCullough, Thomas J., eds. SPACE FOR
MANKIND'S BENEFIT. Proceedings of the first international congress
on "Space for Mankind's Benefit" organized by the Huntsville Association of Technical Societies and held on 15-19 November 1971 at Huntsville, Alabama. NASA SP-313. Washington, D.C.: National Aeronautics and Space Administration, 1972. vi, 479 p. Illus., refs.,
figs. Available from NTIS N73-13829.

   The nine scientific papers published in this volume are divided into ten sessions. Sessions of special interest to remote sensing of the earth environment include session 2 (fundamental benefits of the space program), session 3 (benefits
   of orbital surveys and space technology to environmental protection), and session 4 (earth resources observations through
   orbital surveys).

201.   White, Dennis, ed. RESOURCE SATELLITES AND REMOTE AIRBORNE
SENSING FOR CANADA (PROCEEDINGS OF THE FIRST CANADIAN
SYMPOSIUM ON REMOTE SENSING). 2 vols. Symposium in February 1972 in Ottawa. Ottawa: Canada Centre for Remote Sensing,
1972. Maps, refs., figs., tables.

   Because this was prior to the launch of ERTS-1 in July 1972
   only two of the sixty-eight papers actually deal with satellite data, and these are applications of weather satellite
   imagery. Five papers do, however, discuss ERTS and its potential applications. Forty-six papers deal with aircraft data,
   and of these thirty-four are concerned with aerial photography.

In total, this symposium and its proceedings present a valid overview of the state of remote sensing prior to the advent of the ERTS-1 data.

202. Woll, P.W., and Fischer, William A., eds. PROCEEDINGS OF THE FIRST ANNUAL WILLIAM T. PECORA MEMORIAL SYMPOSIUM. U.S. Geological Survey Professional Paper 1015. Washington, D.C.: U.S. Government Printing Office, 1977. xxxv, 370 p. Illus., refs., figs., tables. Paper.

This is the first of a series of symposia which bear the name of one of the individuals responsible for the development of remote sensing within the geological survey and within the United States. This meeting, held in October 1975 at the EROS Data Center in Sioux Falls, S.Dak., was sponsored by the American Mining Congress in concert with several government and professional organizations. A total of thirty-seven papers covering a range of topics, but especially strong in geologic and cartographic topics, and also encompassing a wide variety of remote sensors (with Landsat being especially visible) are included. More recent symposia, for which the proceedings have not been published to date, include the following:

SYMPOSIUM	DATE	LOCATION
Second	25-29 Oct. 1976	Sioux Falls, S.Dak.

SPONSORS
American Society of Photogrammetry and U.S. Geological Survey (see American Society of Photogrammetry, citation no. 141A)

Third	30 Oct. 2 Nov. 1977	Sioux Falls, S.Dak.

SPONSORS
American Association of Petroleum Geologists and U.S. Geological Survey

203. Zirkind, Ralph. PROCEEDINGS OF THE SYMPOSIUM ON ELECTROMAGNETIC SENSING OF THE EARTH FROM SATELLITES. Symposium held at Coral Gables, Florida, 22-24 November 1965. Brooklyn: Polytechnic Press of the Polytechnic Institute of Brooklyn, 1967. vi, 384 p. Figs.

A total of twenty-four papers and seven abstracts which were presented. The papers may be roughly divided into three (uneven) groups: instrumentation, electromagnetic properties of the earth, and the combination of the two for earth observations. Many of the concepts and ideas presented have since been incorporated into experimental and operational satellites.

Chapter 3
MANUALS AND GUIDES

As is rather obvious from the proceeding sections of the bibliography, much of remote sensing is quite new (indeed, the term was coined only in 1961). Thus, the youth of the "discipline" and the rapidity with which the developments in instruments, techniques, and methodology are progressing dictate that there is a great need for various manuals, instructional aids, information guides, and similar literature to aid the technician and the remote sensing data user to keep abreast of the developments. For, although numerous journal articles, symposia papers, and books give both detailed and general views of the progress, few of them show the time and effort that have been given to yield detailed views of how these results were actually obtained.

Hence, in this section of the bibliography, we have presented the materials concerned with the operation of various computer systems, the processing techniques used for preparation of remotely sensed data and the generally available lesson guides and workshop notes which have been prepared for both introductory and advanced sessions dealing with the use of remotely sensed data. Some manuals have been prepared for very specific audiences, and this will be readily apparent from the introductory statements if not from the titles. Others present a more general view of remote sensing and will deal specifically with a particular data type, data acquisition, or processing system. As before, items which are not generally available have been discarded as being of minimal utility for the readers of this bibliography because of the difficulties in obtaining the literature.

204. Avery, T. Eugene. PHOTO INTERPRETATION FOR LAND MANAGERS. Publication M-76. Rochester, N.Y.: Eastman Kodak Co., 1970. 26 p. Illus., refs., figs., col. and b/w photos. Paper.

> This brief, but excellent, booklet introduces the reader to the use of aerial photographs for interpretation. A short section on the use of Kodak film for aerial photography removes some of the often-perceived mysteries of film. A discussion of several applications, is most valuable. Also included is a short discussion concerning the acquisition of aerial photographs. Fourteen references complete the booklet.

205. Baker, Simon, and Dill, Henry W., Jr., comps. THE LOOK OF OUR
LAND: AN AIR PHOTO ATLAS OF THE RURAL UNITED STATES. 5
vols. Agricultural Handbook nos. 372, 384, 406, 409, and 419.
Washington, D.C.: Economic Research Service, U.S. Department of
Agriculture, 1970–71. Illus., figs. Paper. For sale by U.S. Gov-
ernment Printing Office.

> These atlases bring together text and photographs that show
> land use and related information according to an established
> regional and area classification of U.S. land resources.
> Stereo photographs and a description of the area included
> in the photograph pairs, present a valuable background for
> using such remotely sensed data for mapping and other activ-
> ities. The individual titles in this series are as follows:
> THE FAR WEST, NORTH CENTRAL AND LAKE STATES, THE
> EAST AND SOUTH, THE MOUNTAINS AND DESERTS, and
> THE PLAINS AND PRAIRIES.

206. Bradford, W.R., ed. HANDBOOK OF REMOTE SENSING TECH-
NIQUES. Orpington, Kent, Engl.: Technology Report Centre. Prepared by EMI Electronics for the Department of Trade and Industry
under Contract K46A/59. April 1973. 419 p. Illus., refs. Looseleaf.

> The purpose of this handbook is to provide an authoritative
> summary account of the latest knowledge of the technical
> aspects of remote sensing of the earth from space as applied
> to the study of earth resources and environment. Most as-
> pects of the subject have been covered, including sensors,
> facilities, sites, data transmission, data handling, and in-
> terpretation. An excellent handbook.

207. California, University of Berkeley. REMOTE SENSING OF ENVIRON-
MENT--WORKSHOP. Sponsored by the Association of American Geographers on 27 August 1970. Berkeley, Calif.: August 1970. Var.
pag. Figs., illus., refs. Paper, spiralbound.

> The papers contained in this syllabus have been assembled
> in introduce the field of remote sensing at the basic level.
> These papers were used for a one-day workshop.

208. Colwell, Robert N., ed. MANUAL OF PHOTOGRAPHIC INTERPRE-
TATION. Falls Church, Va.: American Society of Photogrammetry,
1960. xv, 868 p. Figs., tables, illus., bibliog.

> This manual is a comprehensive work describing the principles
> and applications of photo interpretation in various earth sci-
> ences and earth resources fields. It was prepared by one
> hundred recognized authorities and has been thoroughly cross-
> referenced and indexed between the various chapters. At
> the time of its publication it satisfied the need for an

authoritative, comprehensive reference work on photo interpretation and it remains one of the primary reference books
in this field. For a description of the companion version of
this volume, see item 228, MANUAL OF REMOTE SENS-
ING.

209. Corian, Edward F., comp. OPERATIONAL PRODUCTS FROM ITOS
SCANNING RADIOMETER DATA. Washington, D.C.: U.S. National
Environmental Satellite Service, October 1973. 61 p. Refs. Avail-
able from NTIS, N74-10813.

A manual is presented for the potential user of processed
data obtained from the ITOS scanning radiometers. The in-
tent has been to convey primary information in the main
text so as to provide an overview for the customer whether
his mission be oriented toward operations or research. With
this in mind, an attempt is made to minimize the use of
specialized phraseology. It is to be expected that a potential user may thereby assess the potential value of the information for his mission even though his familiarity with spacecraft operations or automatic data processing may be limited.
Specifically, insight may be gained for earth location accuracy, precision and stability of calibrated data, areal extent, and the availability of products.

210. Davis, Shirley M. THE FOCUS SERIES 1975: A COLLECTION OF
SINGLE-CONCEPT REMOTE SENSING EDUCATIONAL MATERIALS.
West Lafayette, Ind.: Laboratory for Applications of Remote Sensing,
Purdue University, 1975. 47 p. Refs. Available from NTIS, N75-
30634.

The Focus Series has been developed to present basic remote-
sensing concepts in a simple, concise way. Issues currently
available are collected here so that more people may know
of their existence.

211. Deer, Albert J. PHOTOGRAPHY EQUIPMENT AND TECHNIQUES:
A SURVEY OF NASA DEVELOPMENTS. NASA SP-5099. Washington,
D.C.: National Aeronautics and Space Administration, 1972. iv,
182 p. Refs., illus., bibliog. Paper. Available from U.S. Government Printing Office.

The Apollo program has been the most complex exploration
ever attempted by men, requiring extensive research, development, and engineering in most of the sciences. Photography has been used each step of the way to document the
efforts and activities, isolate mistakes, reveal new phenomena, and to record much that cannot be seen by the human
eye. Chapters in this publication cover hand-held systems,
vehicle-mounted cameras, tracking photography, engineering
photography, multispectral photography, earth resources photography, and photographic films.

212.	Dellwig, Louis F.; MacDonald, Harold C.; and Waite, William P., eds. RADAR REMOTE SENSING FOR GEOSCIENTISTS: SHORT COURSE NOTES. Lawrence: Center for Research, University of Kansas, 1972-73. Var. pag. Looseleaf.

An excellent set of notes dealing with radar theory, interpretation, and applications, it consists primarily of reprints of articles presented in the open literature. Several sections, such as those dealing with radar theory were written especially for these notes. The original course was one week long. Unfortunately, the notes are not available on microfiche and the original notes are apparently out of print.

213.	General Electric Co. Space Division. DATA USERS HANDBOOK-- NASA EARTH RESOURCES TECHNOLOGY SATELLITE. General Electric Document 71SD4249. Prepared under Contract NAS5-11320. Washington, D.C.: National Aeronautics and Space Administration, September 1971. 208 p. Figs., illus., glossary. Looseleaf.

This is the basic document which explains the operation of the ERTS-1 satellite, the nature of the data transmission, storage and retrieval system, and how to use the computer compatible tapes. Numerous revisions have been published and distributed to the holders of the handbook. This publication is now somewhat dated and superceded by similar documents and by changes in the satellite parameters, but it still provides a good source of general information and background about the technology of the ERTS-1 system.

214.	Higham, A.D.; Wilkinson, B.; and Kahn, D. MULTISPECTRAL SCANNING SYSTEMS AND THEIR POTENTIAL APPLICATION TO EARTH RESOURCE SURVEY. 7 vols. Under contract to ESRO. Havant, Hampshire, Engl.: Plessey Radar Research Centre, South Leith Park House, 1972-73. Figs., refs., illus. Available from NTIS.

The studies that gave rise to this series of reports considered the characteristics of multispectral scanning systems and the applications for which they might be used. The result of reading these reports should be an understanding of the techniques and processes of MSS together with the ability to identify potential applications in the field of interest to the reader. The form of the reports is such that most of the information necessary to understand the principles is presented in a descriptive manner with more complex quantitative factors largely given in the form of appendixes. The series consists of seven volumes:

Vol. 1.	BASIC PHYSICS AND TECHNOLOGY	Dec. 1972
201 p.		N73-22403

Vol. 2.	SPECTRAL PROPERTIES OF MATERIALS	Dec. 1972
257 p.		N73-22404

Vol. 3.	DATA PROCESSING		April 1973
	271 p.	N73-30356	
Vol. 4.	EARTH SCIENCE APPLICATIONS		April 1973
	209 p.	N73-28450	
Vol. 5.	RECOMMENDATIONS FOR EUROPEAN PROGRAMME		March 1973
	119 p.	N73-32301	
Vol. 6.	SUMMARY		June 1973
	215 p.	N74-10385	
Vol. 7.	BIBLIOGRAPHY		June 1973
	82 p.	N73-32299	

215. Horvath, Robert. INTERPRETATION MANUAL FOR THE AIRBORNE REMOTE SENSOR SYSTEM. Ann Arbor: University of Michigan, Environmental Research Institute of Michigan, January 1974. 97 p. Figs., illus. Paper.

The primary purpose of this manual is to provide technical guidelines for those personnel charged with operating and interpreting the output from the Airborne Remote Sensor System. The emphasis in this manual is placed entirely upon the oil-pollution display, detection, and monitoring problem. The system provides the operator with a real time display output of both the infrared and the ultraviolet data. Although the interpretation manual deals with only a limited portion of the EM spectrum and was written for very specific problems, it is a useful guide to the general problem of oil slick detection and monitoring, and gives a good example of one type of interpretation manual which can be prepared.

216. International Astronautical Federation. Working Group One. Committee on Application Satellites. GROUND SYSTEMS FOR RECEIVING, ANALYZING AND DISSEMINATING EARTH RESOURCES SATELLITE DATA. Paris: International Astronautical Federation, November 1974. 95 p. Figs., charts, tables. Available from the American Institute of Aeronautics and Astronautics, New York.

This publication was prepared primarily to explain the availability of earth-resources satellite data and to discuss the ground facilities needed for reception, processing, and dissemination of the available data. Thus this report is essentially a comprehensive survey of the state-of-the-art of such ground systems. The primary emphasis is on ERTS data with examples from Canada and Brazil, together with some general statements concerning economies of the various modes of operation.

217. International Remote Sensing Institute. REMOTE SENSING, 1969. 2 vols. Sacramento, Calif.: International Remote Sensing Institute, 1969. 450 p., 19 p. Paper.

This set of publications, published in support of the Remote Sensing Conference held at Sacramento in 1969, is divided into two major volumes. Volume 1 deals with the general problems of remote sensing (e.g., the limitations, uses, ground truth acquisition, and systems approach) as well as with some specific instrument techniques which are used (e.g., thermal IR, deep probing methods, microwave scatterometers, and multispectral data gathering). Volume 2, originally published by Cartwright Aerial Surveys as their proceedings from the Third Annual Conference on Remote Sensing of Air and Water Pollution, deals primarily with some applications of remote sensing for these pollution problems. Unfortunately, this excellent set of papers and notes is very difficult to locate.

218.   Leberl, Franz. "Radargrammetry For Image Interpreters." Lectures from course held at Bogota, Colombia, from 13 November to 7 December 1972. Enschede, Netherlands: International Institute for Aerial Survey and Earth Sciences, 1973. 136 p. Illus., figs., bibliog. Mimeographed. Spiralbound.

This report is the result of a series of lectures on radargrammetry given at a short course on the use of side-looking airborne radar (SLAR) in the earth sciences. The main point lies in the geometric description of SLAR, the possible error sources, estimates of their effects on the imagery, and the explanation of a number of methods of carrying out geometric measurements in the imagery. The report falls into two main parts. Chapters 2 through 6 analyze the geometry of SLAR. Chapters 7 and 8 concern the use of SLAR to obtain geometric measurements for both individual and overlapping images. An introductory section describes the SLAR system and a concluding chapter deals with a number of radargrammetric considerations on flight planning and evaluation.

219.   Lindenlaub, John C., and Lube, Bruce M. MATRIX OF EDUCATIONAL AND TRAINING MATERIALS IN REMOTE SENSING. Information Note 052576. West Lafayette, Ind.: Laboratory for Applications of Remote Sensing, Purdue University, 1976. 46 p. Available from NTIS, N76-30635.

Remote-sensing educational and training materials developed by LARS have been organized in a matrix format. Each row in the matrix represents a subject area in remote sensing and the columns represent different types of instructional materials. This format has been proved to be useful for displaying in a concise manner the subject matter content, prerequisite requirements, and "technical depth" of each instructional module in the matrix. A general description of the matrix is followed by three examples designed to illustrate how the matrix can be used to synthesize training programs tailored to meet the needs of individual students. A detailed description of each of the modules in the matrix is contained in a "catalog" section.

220.   Lindenlaub, John C., and Russell, James. AN INTRODUCTION TO
        QUANTITATIVE REMOTE SENSING. LARS Information Note 110474.
        West Lafayette, Ind.: Laboratory for Applications of Remote Sensing,
        Purdue University, 1974. ii, 63 p. Available from NTIS, N75-26477.

        This booklet discusses the quantitative approach to remote
        sensing and provides an introduction to the analysis of
        remote-sensing data. It stresses the application of pattern
        recognition in numerically oriented remote-sensing systems.
        Although the specific purpose is to provide a common back-
        ground and orientation for users of the LARS computer soft-
        ware system, it should prove to be helpful for anyone inter-
        ested in introductory readings. The booklet follows the for-
        mat of a programmed text.

221.   Michigan, University of. Engineering Summer Conference. FUNDA-
        MENTALS OF REMOTE SENSING. Course held from 14–25 July 1969.
        Ann Arbor: 1969. Var. pag. Figs., illus., refs. Paper.

        These copyrighted notes cover the field of remote sensing and
        include non–EM remote sensing. The basics of the several
        instruments used in remote sensing are discussed and explained,
        as are the methods of analysis of the resulting data as well
        as comments concerning comparative analysis and the eco-
        nomics of remote sensing.

222.   Michigan, University of. Engineering Summer Conference. PRINCI-
        PLES OF IMAGING RADARS: AN INTENSIVE SHORT COURSE.
        Course held from 22 July to 2 August 1968, 21 July to 1 August 1969,
        and 30 July to 31 August 1970. Ann Arbor: 1968–70. Var. pag.
        Figs., illus., refs. Paper.

        These copyrighted notes which were prepared for limited dis-
        tribution for critical and experimental use only are especially
        useful in providing engineering backgrounds to the principles
        of imaging radars. Included are the introduction to imaging
        radars, synthetic aperture radars, optical processing, holog-
        raphy, data recording, SAR signal, image and SNR, range-
        doppler radars, phase errors, motion compensation, and ap-
        plications of imaging radars, among other subjects.

223.   Michigan, University of. Institute of Science and Technology. Wil-
        low Run Laboratories. THE UNIVERSITY OF MICHIGAN NOTES FOR
        A PROGRAM OF STUDY IN REMOTE SENSING OF EARTH RESOURCES.
        Program conducted at the NASA, MSC in Houston, Texas, 14 February–
        3 May 1968. Report no. 1672-1-X. Ann Arbor: November
        1968. Var. pag. Figs., illus., refs. Paper.

        The notes consist of seven major sections: (1) fundamentals
        and physical principles, (2) radiometry for remote sensors,
        (3) photography, (4) television and magnetic recording, (5)

optical-mechanical scanners, (6) radar and passive microwave radiometry, and (7) data handling and processing. Although a decade old, these notes still are extremely useful for providing background information concerning the hardware and engineering areas of remote sensing. Little attention is given to detailed applications to earth resources. (See citation no. 177).

224.   Miller, Victor C., and Miller, Calvin F. PHOTOGEOLOGY. International Series in the Earth Sciences. New York: McGraw-Hill Book Co., 1961. vii, 248 p. Figs., illus., bibliog., index.

This publication, dealing exclusively with the use of aerial photography for geology and geomorphology studies, is divided into three major parts: mechanics (the collection of aerial photographs and the instruments used in their interpretation), principles of interpretation, and illustrations and exercises. An excellent book of considerable utility for introductory purposes as well as application uses for senior personnel. It contains a list of sources for illustrative material used.

225.   Nunnally, Nelson R. INTRODUCTION TO REMOTE SENSING: THE PHYSICS OF ELECTROMAGNETIC RADIATION. Prepared by the Association of American Geographers in support of the NSF 1969 Summer Short Course in Geographic Applications of Remote Sensing, funded by a U.S. Geological Survey under contract (14-08-0001-10921). Johnson City: East Tennessee State University, April 1969. 39 p. Illus., refs. Paper.

This small paper is very good introductory statement concerning the physics of EM radiation.

226.   Raytheon Co. Autometric Operation. TRAINING COURSE ON DATA REDUCTION OF RADAR TOPOGRAPHIC IMAGERY. Alexandria, Va.: February 1969. Var. pag. Available from NTIS, AD-721 653.

This basic training course is an excellent introduction to radar remote-sensing interpretation. Basic concepts of radar operation, techniques, and procedures for planning radar imaging missions, imagery examples of malfunctions, and techniques of radargrammetry are all discussed in detail.

227.   Reeves, Robert G., ed. AN INTRODUCTION TO ELECTROMAGNETIC REMOTE SENSING WITH EMPHASIS ON APPLICATIONS TO GEOLOGY AND HYDROLOGY. Course held on 7-9 November 1968 at Houston, Tex. Washington, D.C.: American Geological Institute, 1968. Var. pag. Illus., maps. Paper.

This book consists of the AGI Short Course lecture notes for the course. As such, they are a little "rough" but provide an excellent background for basic remote sensing. Note the early publication date, however.

228. Reeves, Robert G., et al., eds. MANUAL OF REMOTE SENSING.
2 vols. Falls Church, Va.: American Society of Photogrammetry,
1975. Figs., tables, plates, bibliog., refs.

This manual was written with two major objectives: (1) to
replace and update the material contained in the MANUAL
OF PHOTOGRAPHIC INTERPRETATION (see item 208) and
(2) to provide information on the new techniques of remote
sensing and their uses. All aspects of the subject are covered, and like the earlier manuals prepared by the ASP,
this manual provides the single best, concise source of information on the subject.

229. Smith, John T., Jr., et al., eds. MANUAL OF COLOR AERIAL
PHOTOGRAPHY. Falls Church, Va.: American Society of Photogrammetry, 1968. xv, 550 p. Figs., tables, illus., col. plates., refs.

This is the most complete and readily available source and
resource book concerning aerial photography using color
films. Information concerning the nature of color, mission
planning, films, chemistry of emulsions, printing techniques,
metric quality of color aerial films, color photography from
space, and the photographic interpretation of color films
(with numerous excellently reproduced examples) are included
in the publication. Although almost a decade old, only the
technical material concerning films is obviously dated. The
other sections retain the importance they had when first published.

230. Strandberg, Carl H. AERIAL DISCOVERY MANUAL. Wiley Series on
Photographic Science and Technology and the Graphic Arts. New
York: J. Wiley and Sons, 1967. xiii, 249 p. Figs., illus., refs.
Paper (spiralbound).

This manual has been designed for classroom use in aerial
photo interpretation and is divided into three major sections:
aerial photographic interpretation, photogeology, and photo-
hydrology. Numerous excellent examples of applications and
stereo pairs of photography have ensured that this publication
will retain its premier position for years to come. Recently
(1975) a Spanish edition translated by David S. Congost has
been prepared.

231. Thompson, Morris M., et al., eds. MANUAL OF PHOTOGRAMMETRY.
3d ed. 2 vols. Falls Church, Va.: American Society of Photogrammetry, 1966. Figs., illus., refs.

An extensive revision of the second edition (1952) of this
manual comprises this standard reference book. A total of
twenty-five chapters covering all aspects of photogrammetry
is included. An excellent and necessary research and resource book.

232. U.S. Geological Survey. SYLLABUS: WORKSHOP ON REMOTE
      SENSING AND ERTS IMAGE INTERPRETATION, held at the EROS
      Data Center, Sioux Falls, S.Dak., from 30 May to 28 June, 1974.
      USGS Open File Report 75-196. Washington, D.C.: June 1974.
      xiii, var. pag. Available from NTIS, N76-14585, or PB 243 933.

      This is the syllabus of a training course which had major
      objectives: (1) to train engineers, scientists, and managers
      involved in resources investigations of the use of remotely
      sensed data, especially that provided by ERTS-1 and similar
      satellites, and (2) to demonstrate the value of remote sensing
      to interdisciplinary/multidisciplinary cooperation in environ-
      mental analyses, resource inventories, and land use planning.
      Fourteen major sections comprise the syllabus including de-
      scriptions of the satellites and data, preparation of the data
      into mosaics and/interpretation principles as applied to the
      several earth science oriented disciplines. Four appendixes
      are included.

233. U.S. National Aeronautics and Space Administration. Educational
      Programs Division. Petrillo, Anthony J., director, Jefferson County,
      Colo. Public Schools Interdisciplinary Environmental Education Team.
      NASA EP-103. WHAT'S THE USE OF LAND: A SECONDARY
      SCHOOL SOCIAL STUDIES PROJECT. Washington, D.C.: U.S.
      Government Printing Office, October 1976. 57 p.

      This document, prepared as a part of the ongoing educational
      activity of NASA, shows how various types of remotely sensed
      imagery can be combined with information from several other
      sources to aid in teaching several school curriculum topics.
      It also provides an example of the value of using information
      from many sources in the process of evaluating events and
      making decisions. The publication is divided into three ma-
      jor sections following the introduction and statement of ob-
      jectives: (1) "Land Use Studies," describing the multidis-
      ciplinary unit concept; (2) "Types of Data for Land Use
      Studies", which gives advice on where to obtain data for
      surveys and how to use those data, and (3) "Appendix,"
      which discusses the different factors that influence the use
      of land.

234. U.S. National Aeronautics and Space Administration. Goddard Space
      Flight Center. THE NIMBUS 6 USER'S GUIDE. Greenbelt, Md.:
      February 1975. xv, 227 p. Paper.

      This document provides potential data users with background
      information on the Nimbus 6 spacecraft and experiments as
      a basis for selecting, obtaining, and utilizing Nimbus 6 data
      in research studies. The basic spacecraft system operation
      and the objectives of the Nimbus 6 flight are outlined, fol-
      lowed by a detailed discussion of each of the experiments.
      The format, storage, and access to the data are also described.

Finally, the contents and format of the NIMBUS 6 DATA CATALOGS are described (see item no. 266). These catalogs were issued periodically after the launch of Nimbus 6. They contained representative pictorial data and daily THIR montages obtained during each period, as well as information on the collection and availability of all Nimbus 6 data.

235. U.S. National Aeronautics and Space Administration. Johnson Space Center. EARTH RESOURCES PROGRAM: PROCEDURES MANUAL FOR DETECTION AND LOCATION OF SURFACE WATER USING ERTS-1 MULTISPECTRAL SCANNER DATA. 5 vols. Houston: December 1973. Available from NTIS.

A computer-aided procedure, for use in the detection and location of areas of surface water, has been developed in support of the National Program of Inspection of Dams established by Public Law 92-367. The procedure utilizes data acquired by the unmanned Earth Resources Technology Satellite (ERTS-1) in conjunction with ancillary data in the form of topographic and highway maps, and meteorological data summaries. The procedure is divided into a five-volume set as follows:

Vol.				
1.	SUMMARY	December 1973	54 p.	N74-32795
2.	DATA ACQUISI-TION	November 1973	34 p.	N74-32796
3.	CONTROL NET-WORK ESTAB-LISHMENT	November 1973	53 p.	N74-32797
4.	COMPUTER PRO-GRAM DE-SCRIPTION AND USERS GUIDE	November 1973	233 p.	N74-32798
5.	INFORMATION CORRELATION AND INTER-PRETATION	November 1973	41 p.	N74-32799

Much of the work was done under contract by Lockheed Electronics Co., Inc.

236. U.S. National Aeronautics and Space Administration. Manned Spacecraft Center. EREP USERS HANDBOOK. Rev. ed. Houston: March 1971. 379 p. Refs. Available from NTIS, N72-13851.

Revised Skylab spacecraft, experiments, and mission planning information is presented for the Earth Resources Experiments Package (EREP) users. The major hardware elements as well as the medical, scientific, engineering, technology, and earth resources experiments are described. Ground truth

measurements and EREP data handling procedures are discussed.
The mission profile, flight planning, crew activities, and
aircraft support are also outlined.

237. Von Frijtag Drabbe, C.A.J. AERIAL PHOTOGRAPH, AND PHOTO
INTERPRETATION. Amsterdam: J.H. de Bussy, n.d. 80 p.

This brief manual, is an introductory statement to the interpretation of aerial photographs primarily for geological and
geomorphological studies. The author was the managing director of the Netherlands Topographic Service and the examples deal primarily with postglacial deposits and terrains
from the Netherlands. An excellent example of this type
of work.

238. Wiedel, Joseph W., and Kleckner, Richard. USING REMOTE SENSOR
DATA FOR LAND USE MAPPING AND INVENTORY: A USERS
GUIDE. U.S. Geological Survey Interagency Report 253. Washing-
ton, D.C.: U.S. Geological Survey and Association of American
Geographers, July 1974. 124 p. Figs., illus., refs. Available
from NTIS, N76-14583, or PB-242-813.

This report is an attempt to provide users with various techniques for obtaining and analyzing land use data. The first
step is to acquire remotely sensed data and supplemental
data. Next a coordinate system, plotting base, and land
use classification system are selected, followed by interpretation and delineation of land use. Editing procedures, including field checking, validate the accuracy of the mapping. Application of automated techniques for computer-
generated map products, tabular printouts, area measurements, and temporal aspects of land use are considered.
This is an especially interesting publication in that it presents methods for obtaining various types of remotely sensed
and supplemental data.

239. Wolfe, William L., ed. HANDBOOK OF MILITARY INFRARED TECH-
NOLOGY. Washington, D.C.: Office of Naval Research, Department
of the Navy, 1965. xiii, 906 p. Illus., figs., index. Available
from U.S. Government Printing Office.

This book, comprised of twenty-two chapters, is basically a
compendium of infrared technology, system design, radiation
theory, optics, and processing techniques. Although oriented
toward military applications, the principles and physics discussed, together with the equations presented, are equally
applicable to earth resources studies in this portion of the
EM spectrum.

Chapter 4

CATALOGS

In a recent and excellent small book, Dick Kroeck notes that one of the really big problems is to determine what types of remotely sensed data are available, where these are stored, and how one proceeds to obtain them. If we consider only the aerial photography of the United States, and then only those data which are contained in the federal records and depositories, he estimates that over twenty million photographs are available, with an additional one million being added each year. Add to these aerial photographs the data which are being collected from space and, in the case of weather satellites, have been collected for over a decade, it is clear that some cataloging system and archival retrieval system should be made available. The EROS Data Center, operated by the U.S. Geological Survey, has taken that task into hand and was, indeed, developed specifically for the handling of Landsat data, although its responsibility has been broadened to include other areas and sensors as well.

However, in addition to the "routine" catalogs prepared for ongoing systems on a monthly or annual basis, there are a number of catalogs which deal with specific features of the earth's surface, such as volcanoes or urban land use, or only with a specific remote sensor instrument such as aerial photography or Landsat. An attempt has been made to detect these documents and include them in this present information guide. Most of the catalogs and references listed give information concerning the quality of the data, the times and areas covered, and similar information. Obtaining the data is a somewhat more difficult task, for often such details are not included. Also, this section of the information guide deals only with catalogs and lists; often many publications such as those included in the chapter on general literature in this information guide will contain important ordering information and therefore they should also be closely checked. Dick Kroek, EVERYONE'S SPACE HANDBOOK, Arcata, Calif.: Pilot Rock, 1976 (see no. 249).

240. Barwis, John H. CATALOG OF TIDAL INLET AERIAL PHOTOGRAPHY. Ft. Belvoir, Va.: Coastal Engineering Research Center, U.S. Army Corps of Engineers, June 1975. 179 p. Available from NTIS, N76-13714.

Data on approximately six thousand aerial photographic cov-
erages of tidal inlets are presented in tabular form along
with information on how any given photograph may be obtained. The compilation covers inlets along the Atlantic,
Gulf, and Pacific coasts of the contiguous U.S. coastline
from 1938–74, and includes the following information: inlet name; geographic coordinates; National Ocean Survey
navigation chart covering inlet; georef grid square; month
and year of photograph; federal, state, or commercial agency
holding film; project number; pertinent exposure numbers;
scale; and film type. Information is also given on sources
of additional photography, and on obtaining photography of
beach areas between any two inlets. An index, by U.S.
Army Corps of Engineers district, is also included in this
exceptionally useful catalog.

241. Bird, J.B.; Morrison, A.; and Chown, M.C. WORLD ATLAS OF
PHOTOGRAPHY FROM TIROS SATELLITES I TO IV. Washington,
D.C.: U.S. National Aeronautics and Space Administration, Septem-
ber 1964. iv, 152 p. Illus. Paper.

The major part of this report is a selection of over four hundred photographs taken by TIROS satellites 1 to 4 including
one or more pictures of every state of the United States and
most of the land areas of the world between 55°N Latitude
and 55°S Latitude. They are identified as clearly as the
data permit by distinctive and commonly recognized land-
marks.

242. Denney, Charles S.; Warren, Charles R.; Dow, Donald H.; and Dale,
William J. A DESCRIPTIVE CATALOG OF SELECTED AERIAL PHO-
TOGRAPHS OF GEOLOGIC FEATURES IN THE UNITED STATES.
U.S. Geological Survey. Professional Paper 590. Washington, D.C.:
U.S. Government Printing Office, 1968. 133 p. Figs., plates.

A set of aerial photos selected from U.S. government sources
that illustrate numerous types of geologic features in the
United States comprise this publication. 857 photographs
are grouped in 317 sets including from one to six contact
prints, arranged on a geographic basis by state. A brief
description of the features illustrated and references to the
geologic or topographic maps of the area photographed are
given. Information on how to obtain the various photographs
is provided in this excellent publication.

243. Dismachek, Dennis C., ed. NATIONAL ENVIRONMENTAL SATEL-
LITE SERVICE CATALOG OF PRODUCTS. NOAA Technical Memorandum NESS 88. Washington, D.C.: June 1977. ix, 102 p.

This volume, an update of the CATALOG OF OPERATIONAL
SATELLITE PRODUCTS (see item 247), has been compiled to
provide users with a description of all the currently avail-

able operational products of the National Environmental Satellite Service (NESS). A product is defined as any item routinely produced and available for applications within the environmental sciences, and includes photographic displays, charts, teletype messages, and alphanumeric data.

244. Giddings, L.E. GEMINI PHOTOGRAPHS OF THE WORLD: A COMPLETE INDEX. Houston: Lockheed Electronics Co., May 1975. iv, 101 pp. refs. Available from NTIS N77 27472.

244A. _____. INDEX MAPS FOR GEMINI EARTH PHOTOGRAPHY. Houston: Lockheed Electronics Co., April 1975. 97 p. Refs., maps. Available from NTIS, N75-28502.

Index maps for the Gemini missions are presented; these are for the Gemini 3 through Gemini 12 missions. The maps are divided into four sections: (1) the whole earth, (2) the Western hemisphere and eastern Pacific Ocean, (3) Africa, India, and the Near East, and (4) Asia, Australia, and the Pacific Ocean. The large format of the original copy makes the microfiche copy difficult to read, due to the many fold-out frames. This document presents all of the Gemini index maps known to persons closely attached to the Gemini projects. It is an uncritical listing, published for the convenience of users of Gemini photography.

244B. _____. NEAR EARTH PHOTOGRAPHS FROM THE APOLLO MISSIONS AND THE APOLLO–SOYUZ TEST PROJECT. 4 vols. Houston: Lockheed Electronics Co., August 1977. Available from NTIS, N78-17436 to N78-17439 inclusive.

This is the most authoritative catalog of the near earth photographs of the Apollo Missions and the Apollo–Soyuz Test Project. All photographs are identified as follows: JSC identification number, percent cloud cover, geographical area in sight and miscellaneous information. In addition, details are given on cameras, filters, films, and other technical details. This document contains the primary reference documents for identification of all Apollo and Apollo–Soyuz Test Project photographs of the earth. Index maps are included for Apollo 6, 7, and 9 and the Apollo–Soyuz Test Project. Information on the acquisition of these photographs is also given. No photographs are included in the documents but there is an extensive listing of the scenes by the earth science disciplines.

245. Hassel, Philip G., et al. COLLATION OF EARTH RESOURCES DATA COLLECTED BY ERIM AIRBORNE SENSORS. Ann Arbor: Environmental Research Institute of Michigan, September 1975. 181 p. Illus., figs. Paper. Available from NTIS, N76-10556.

Earth-resources imagery from nine years of data collection with developmental airborne sensors is cataloged for reference. The imaging sensors include single and multiband scanners and side-looking radars. The operating wavelengths of the sensors include ultraviolet, visible and infrared band scanners, and X and L band radar. Imagery from all bands (radar and scanner) were collected at some sites; many sites have repeated coverage. The multiband scanner data were radiometrically calibrated. Illustrations show how the data can be used in earth-resources investigations. References are made to published reports which have made use of the data in completed investigations. Data collection sponsors are identified and a procedure for gaining access to the data is outlined. This document is the best source for both ERIM imagery and documents published by the ERIM staff.

246. Heiken, Grant. CATALOGUE OF SATELLITE PHOTOGRAPHY OF THE ACTIVE VOLCANOES OF THE WORLD. Los Alamos Scientific Laboratory Report LA-6279-MS. Los Alamos, N.Mex.: March 1976. iii, 25 p. Paper. Available from NTIS, N77-18536.

This is a catalog of active volcanoes as viewed from earth-orbiting satellites. The listing was prepared from quality screened photographs selected from the data collected by earth resources technology satellites (ERTS and Landsat), Skylab, Apollo, and Gemini spacecraft. At present, there are photographs of nearly every active volcano in the world. These photographs are particularly useful for regional studies of volcanic fields. Information given for each volcano includes geographical location (by political units), volcano name, image number, and a short general comment concerning the image (i.e., color, size, stereo, amount of coverage of the volcano).

247. Hoope, Eugene R., and Ruiz, Abraham L., eds. CATALOG OF OPERATIONAL SATELLITE PRODUCTS. NOAA Technical Memorandum, NESS 53. Washington, D.C.: U.S. National Environmental Satellite Center, March 1974. ix, 91 p. Illus., figs. Paper. Available from NTIS, N74-25898.

This catalog is designed to acquaint the user community with the products generated from data acquired by sensors carried on environmental satellites controlled by the National Environmental Satellite Service (NESS). A brief description of the system is given; more detailed information can be found in the literature. The emphasis is on operations. A product is defined as any item routinely produced for applications within the environmental services. These range from basic photographic images through a variety of manually-produced and computer-produced interpretation products. These products take the form of facsimile transmissions, photographic images, alphanumeric messages, and digital magnetic tapes. The report contains thirty-nine figures, four tables of acronyms, and an abbreviation list. This volume was updated in 1977 (see item no. 243).

248. International Bank for Reconstruction and Development. Cartographic Division. LANDSAT INDEX ATLAS OF THE DEVELOPING COUNTRIES OF THE WORLD. Washington, D.C.: World Bank; Baltimore: Distributed by Johns Hopkins University Press, February 1976. 17 p. Paper, oversized. Spiralbound.

This atlas is designed to assist developing nations in the acquisition and incorporation of satellite remote sensing into their development planning activities. It displays the satellite coverage available and describes the interpretation techniques. Some nondeveloping countries are included to provide continuity of map coverage. Only Landsat imagery from July 1972 to May 1975 is indexed.

249. Kroeck, Dick. EVERYONE'S SPACE HANDBOOK: A PHOTO IMAGERY SOURCE MANUAL. Arcata, Calif.: Pilot Rock, 1976. 175 p. Illus., refs., bibliog. Paper.

The size of this book grossly understates its importance. It consists of eight chapters: (1) introduction, (2) image characteristics, (3) platforms-sensors-imagery, (4) imagery sources--federal, (5) direct sources--state and private companies, (6) information sources--other data, (7) special data collections, and (8) teacher's section. An extensive listing of sources of data, together with addresses, telephone numbers, and procedures for ordering imagery is provided and makes this an excellent and extremely valuable tool for researchers and teachers.

250. Lockheed Electronics Co. SKYLAB-3 HANDHELD PHOTOGRAPHY ALPHABETIZED GEOGRAPHICAL FEATURE LIST. Houston: Johnson Space Center, National Aeronautics and Space Administration, April 1974. 92 p. Available from NTIS, N74-32804.

The SKYLAB-3 HANDHELD IMAGERY CATALOG is oriented to magazine and frame numbers to facilitate imagery tracking and provides a summary of features visible for each photograph. Imagery researchers require the opposite, that is, given the state, country or other feature name, to find which images contain the given features. The data contained herein has been sorted by geographical name to assist imagery researchers. Most of the names used are of political divisions, no use is made of geographic features (e.g. mountains, volcanoes, salt pans), unless each of these has been given a proper name (e.g., Columbia River).

250A   May, John R.   GUIDANCE FOR APPLICATION OF REMOTE SENSING
        TO ENVIRONMENTAL MANAGEMENT; APPENDIX A: SOURCES OF
        AVAILABLE REMOTE SENSOR IMAGERY.   Instruction Report--M-78-2,
        Appendix A.   Vicksburg, Miss.:   U.S. Army Waterways Experiment
        Station, Mobility and Environmental Systems Laboratory, May 1978.
        32 p.   Available from NTIS; N78-28588.

        > The report, to which this citation is the appendix, was still
        > in preparation as of May 1978.   Three large tables make up
        > the bulk of this report, viz.:   (1) Summary of available re-
        > mote-sensing imagery--Federal Agencies; (2) Summary of
        > available remote-sensing imagery--Corps of Engineers Agen-
        > cies, and (3) Summary of available remote-sensing imagery--
        > State Agencies.   This is a quite useful and current listing
        > of such data.   Photography, both black and white color,
        > and IR dominate the listings.

251.   Mutter, Douglas L., comp.   SPACE, AIR AND PHOTO IMAGES FOR
        THE ROCKY MOUNTAIN STATES.   Technical Report no. 28.   Denver:
        Federation of Rocky Mountain States, July 1973.   86 p.   Illus., maps,
        appendixes.   Paper.   Spiralbound.

        > The purpose of this report is to provide users of remote-
        > sensing data with a starting point for their search for the most
        > appropriate and available imagery which will fit their needs.
        > Satellite as well as high and low altitude aerial photography
        > are covered.   The former includes Gemini, Apollo, Skylab,
        > and ERTS (Landsat) data.   The aerial photography data are
        > from several U.S. government sources.   The Federation of
        > Rocky Mountain States includes Colorado, Idaho, New Mex-
        > ico, Montana, Utah, and Wyoming.   Ordering information
        > for obtaining the remotely sensed imagery is included.

252.   New Mexico, University of.   Technology Applications Center.   APOL-
        LO 9 SYNOPTIC PHOTOGRAPHY CATALOG.   Albuquerque, N.Mex.:
        September 1969.   64 p.   Maps.   Paper.

        > Imagery taken with the hand-held camera (color photography)
        > on Apollo 9 8 to 12 March 1969), and SO-65 multispectral
        > photographs of the same dates are cataloged.   Geographical
        > areas, frame numbers, latitude and longitude, percentage of
        > overlap, altitude of the spacecraft, percentage of cloud cov-
        > er, viewing mode, and general comments describing the lo-
        > cation of the various scenes are included.   No imagery is
        > presented.

253.   _____.   APOLLO 6 AND 7 SYNOPTIC PHOTOGRAPHY CATALOG.
        Albuquerque, N.Mex.:   n.d.   49 p.   Paper.

        > Although no imagery is presented, this publication catalogs
        > all of the photography taken with the 70mm Maurer 220G

camera system on 4 April 1968. Identifications, conducted by R.W. Underwood and H.A. Tiedemann of the Manned Spacecraft Center (now Johnson Space Center) include photo frame number, latitude, longitude, altitude of the spacecraft, percent of clouds on the frame, and the location and description of the areas seen. The same format is also used for the Hasselblad 500-C camera data taken during Apollo 7 of 11-12 October 1968.

254. _____. GEMINI SYNOPTIC PHOTOGRAPHY CATALOG. Albuquerque, N.Mex.: n.d. 36 p. Maps. Paper.

Indexing of these photographs on a series of maps at scales of 1:39,000,000 was by H.A. Tiedemann. Gemini flights 4, 5, 6, 7 are cataloged for the SO-65 experiment. A preliminary index of geographic areas is given, and the catalog consists of frame number, a description of the general geographic area, picture quality, cloud cover, and the degree of angularity. No comments are included about the earth resources of the scenes included in the various frames. No imagery is presented.

255. Richter, Dennis M. "Urban Photo Index for Eastern U.S." PHOTOGRAMMETRIC ENGINEERING 37 (1971): 54-66.

With the increased interest in urban planning as well as industrial site location and analysis it has become increasingly necessary to utilize aerial photography on a sequential basis. The index concerns itself with type locations in terms of urban settings and characteristic industrial plants as they are found in the eastern United States. Eight photographs are included in the paper and a total of 122 are indexed. Index information includes city and county names, photo date, photo identification number(s), and a listing of the USGS quadrangle map covering the photo. All photographs are from the USDA files.

256. Szuwalski, Andre. COASTAL IMAGERY DATA BANK: INTERIM REPORT. U.S. Coastal Engineering Research Center. Miscellaneous Paper no. 3-72. Washington, D.C.: Coastal Engineering Research Center, U.S. Army Corps of Engineers, November 1972. 55 p. Illus. Available from NTIS, N73-22368, and AD 755 508.

This report provides information about the Coastal Imagery Data Bank being compiled by the Coastal Engineering Research Center (CERC). The data bank will consist of a systematic index identifying available aerial photographs of the coastal areas of the United States. Compilation is scheduled for completion in fiscal year 1977. This interim report covers data compiled through fiscal year 1972. Imagery for the index is compiled by the Defense Mapping Agency Topographic

Center (DMATC) under support and direction of CERC. A listing of the available data sources and methods of acquisition of the data is included. The data listed are from the Baltimore, Charleston, Seattle, and Portland U.S. Army Corps of Engineers' districts.

257. Underwood, Richard W. SKYLAB 4: PHOTOGRAPHIC INDEX AND SCENE IDENTIFICATION. Report NASA-TM-X-72440. Houston: Johnson Space Center, National Aeronautics and Space Administration, June 1974. Illus., figs. 335 p. Available from NTIS, N75-27536.

This document has been prepared as a "quick reference guide" to the photographic imagery obtained on Skylab 4. Place names and descriptors used give sufficient information to identify frames for discussion purposes and are not intended to be used for ground nadir or geographic coverage purposes. Frames are identified by camera used (70mm Hasselblad, 35mm Nikon, 70mm 6-lens multispectral camera, etc.) and by number. A cross index of frame number on each roll of film and the NASA number are provided. Film types, lens length and so forth are included. No exposure information is given. Photographs taken between 16 November 1973 and 8 February 1974 are included in this index.

258. Underwood, Richard W., and Holland, John W. SKYLAB 3: PHOTO-GRAPHIC INDEX AND SCENE IDENTIFICATION. Houston: Johnson Space Center, National Aeronautics and Space Administration, November 1973. 257 p. Paper. Available from NTIS, N74-16061.

This document has been prepared as a "quick reference guide" to the photographic imagery obtained on Skylab 3. Place names and descriptors used give sufficient information to identify frames for discussion purposes and are not intended to be used for ground nadir or geographic coverage purposes. Data included are from the following systems: 70mm Hasselblad (earth looking), 35mm Nikon (earth looking, 70mm, 6-lens multispectral camera, 18 inch (450mm) focal length earth terrain camera, and 16mm motion picture camera. No maps or figures are included to indicate locations of the data.

259. U.S. Geological Survey. AVAILABILITY OF EARTH RESOURCES DATA. USGS INF-74-30. Washington, D.C.: U.S. Government Printing Office, 1976. 46 p. Paper.

This small booklet is quite necessary for any remote-sensing data collection efforts as it lists addresses and products of the several organizations which offer remotely sensed data for use by the scientists and general public. The purpose of this booklet is to help the reader to become aware of the volume of earth-resources data that have been collected by agencies of the U.S. government and to find out where and how these data may be obtained. Aerial photographs, multispectral scanners, radar, and other information collected from both spacecraft and aircraft are covered. A most useful publication.

260. _____. INDEX TO LANDSAT COVERAGE. WPS-1--WPS-7. Reston, Va.: 1975. 7 col. maps. Scale varies.

This set of seven index maps summarize nominal scene coverage of Landsat-1 images received and processed by NASA for the period of 23 July 1972 to 23 July 1974. Symbols indicate the least amount of cloud coverage for a nominal scene. A nominal scene is a geographic location over which repetitive images are centered when the satellite maintains normal positional tolerances. All scenes are identified by orbit path and row number. The sheets available in this series are as follows:

WPS-1	EURASIA	1:18,000,000
2	NORTH AMERICA	1:18,000,000
3	SOUTH AMERICA	1:18,000,000
4	AFRICA	1:18,000,000
5	OCEANIA	1:18,000,000
6	ANTARCTICA	1:18,000,000
7	ARCTIC OCEAN	1:9,000,000

261. U.S. National Aeronautics and Space Administration. Ames Research Center. Airborne Mission and Applications Division. AIRBORNE INSTRUMENTATION RESEARCH PROJECT, SUMMARY CATALOGS. 10 vols. Moffett Field, Calif.: 1973. Additional information is available from AIRP Data Facility, Mail Stop 211-8, NASA Ames Research Center, Moffett Field, California 94035. Maps. Paper.

This series of catalogs is published to describe the data collected by the Earth Resources Aircraft Project (ERAP, now Airborne Instrumentation Research Project, AIRP). This project operates two U-2 high altitude aircraft in support of NASA's Earth Observations Program. The catalog is assembled from the key elements of the "Flight Summary Reports published for each data collection flight conducted by the project. These elements are a data summary, a flight summary, and map showing ground tracks of the aircraft during data collection. The summary catalogs are serial, represent the collection of data flights, and are in chronological order.

262. U.S. National Aeronautics and Space Administration. Audio Visual Branch. Public Information Division. SPACE PHOTOGRAPHY 1977 INDEX. Washington, D.C.: [1978?] 198 p. Available at no charge from Audio Visual Branch, Public Information Division Code FP, NASA, 400 Maryland Ave., SW, Washington, D.C. 20546.

This book is intended to be an index of representative space photographs available from NASA. The transparencies which are listed are available on loan, free to the media, and at a nominal laboratory charge for other users. An excellent chronology of the launches of satellites (together with dates, mission descriptions, etc.) is included together with the listings of those photographs which are available. Many of the available photographs are concerned with the astronauts and their activities, but there is an extensive listing of Gemini terrain photographs and the NASA high altitude aircraft infrared photographs.

263. U.S. National Aeronautics and Space Administration. Goddard Space Flight Center. LANDSAT NON-U.S. STANDARD CATALOG. Greenbelt, Md.: 1972-- . Paper. Available from NTIS.

This series of catalogs supplements those issued under the title LANDSAT U.S. STANDARD CATALOG (item no. 264) and covers those portions of the world not covered by that catalog. Areas covered by a Landsat frame which could be included in either catalog are normally contained in the LANDSAT U.S. STANDARD CATALOG. See item no. 264 for a detailed description of the contents of this publication.

264. _____. LANDSAT U.S. STANDARD CATALOG. Greenbelt, Md.: 1972-- . Paper. Available from NTIS.

This series of catalogs, published by the Goddard Space Flight Center, is prepared to provided dissemination of information regarding the availability of Landsat imagery. Thus, the catalogs are published on a monthly schedule and identify imagery which has been processed and put into the data files during the referenced month. This catalog includes imagery covering the continental United States, Alaska, and Hawaii. In addition, as a supplement to these catalogs, Landsat imagery of one spectral band is available on 16mm film. In addition to the routine monthly catalog, a cumulative catalog for each Landsat satellite, covering a year based on the launch data for that satellite is also published. The data included in these catalogs are satellite coverage maps, observation identification number listings, longitude and latitude listings, and numerous types of information concerning the type and quality of the data available. These catalogs are supplemented by the LANDSAT NON-U.S. STANDARD CATALOG (item no. 263) and all are listed in STAR (item no. 364).

265. _____. NIMBUS 5 DATA CATALOG. 12 vols. Volumes 1 through 9 were prepared by Allied Research Associates, Baltimore, Md., under Contract NAS 5-21617; volumes 10 through 12 were prepared by Management and Technical Services Co. under Contract NAS 5-20694.

Greenbelt, Md.: 1973-75. Paper.

This series of catalogs was published by NASA to document data acquired from the Nimbus 5 Meteorological Satellite and contain documentation for succeeding calendar months throughout the useful lifetime of Nimbus 5. The Nimbus 5 catalog presents the type of data available, anomalies in the data, if any, and geographical location and time of data. In addition, the first volume presents some preliminary results from various Nimbus 5 experiments. Background information concerning the Nimbus 5 Meteorological Satellite system and a description of the data formats have been published separately in the NIMBUS 5 USER'S GUIDE, with postlaunch USER'S GUIDE information changes and corrections included in the first volume of the DATA CATALOG.

		DATES OF COVERAGE	ORBITS	PUB. DATE
Vol.	1	19 December 1972 21 January 1973	104-693	May 1973
	2	1 February 1973 31 March 1973	694-1485	September 1973
	3	1 April 1973 31 May 1973	1486-2304	December 1973
	4	1 June 1973 31 July 1973	2305-3123	February 1974
	5	1 August 1973 30 September 1973	3124-3942	March 1974
	6	1 October 1973 30 November 1973	3943-4761	April 1974
	7	1 December 1973 31 January 1974	4762-5593	June 1974
	8	1 February 1974 31 March 1974	5594-6385	July 1974
	9	1 April 1974 31 May 1974	6386-7204	August 1974
	10	1 June 1974 31 July 1974	7205-8023	October 1974
	11	1 August 1974 30 September 1974	8024-8842	April 1975
	12	1 October 1974 30 November 1974	8843-9660	August 1975

266. _____. NIMBUS 6 DATA CATALOG. 3 vols. Prepared by the Management and Technical Services Co. under Contract NAS 5-20694. Greenbelt, Md.: 1975-76. Paper.

This series of catalogs was published by NASA to document data acquired from the Nimbus 6 meteorological satellite. The first volume covers the period from 12 June 1975 through 31 August 1975 (orbits 1-1082), the second volume covers the period from 1 September 1975 through 31 October 1975 (orbits 1083-1900), and the third volume covers the period 1

November 1975 through 31 December 1975 (orbits 1901–
2717). Subsequent catalogs will contain documentation for
succeeding periods throughout the useful lifetime of Nimbus
6. Background information concerning the Nimbus 6 meteorological satellite system and a description of the experiments and data formats has been published separately in the
NIMBUS 6 USER'S GUIDE (item no. 234). Postlaunch
USER'S GUIDE information changes and corrections are included in the data catalogs. The Nimbus 6 catalogs present the type of data available, anomalies in the data, if
any, and geographic location and time of the data.

267.   U.S. National Aeronautics and Space Administration. Johnson Space
Center. SKYLAB EARTH RESOURCES DATA CATALOG. Prepared in
cooperation with Martin Marietta, Denver. Houston: 1974. 393 p.
Refs. Available from NTIS, N75-20798.

An index of Earth Resources Experiment Program photographs
is provided along with information on how these data can
be obtained. Suggestions are presented for possible utilization of the data in the following areas: land resources management, water resources, marine resources, landform surveys, geologic mapping and mineral resources, agriculture,
forest and range resources, and environmental applications.
It is intended to stimulate potential users to apply the data
to their respective fields of interest. Extensive descriptions
of the instruments and the entire Skylab mission are also included. Original contains color imagery.

268.   U.S. National Aeronautics and Space Administration. Manned Space-
craft Center. EARTH RESOURCES RESEARCH DATA FACILITY INDEX.
Rev. ed. Houston: July 1970. 419 p. Refs. Available from NTIS,
N71-12199.

The cumulative issue of the EARTH RESOURCES RESEARCH
DATA FACILITY INDEX is presented. All Earth Resources
Program information and related data that are available at
the NASA-MSC are listed. Included in the index are data
collected during flights over test sites and from studies made
by investigators supporting the Earth Resources Survey Program. The information is cataloged into four main categories: (1) technical documents and maps, (2) functional and
checkout data, (3) imagery data, and (4) electronic data.

269.   _____. EARTH RESOURCES RESEARCH DATA FACILITY R&D FILE.
2 vols. Houston: July 1971. 505 p. 339 p. Paper. Available
from NTIS, N72-13302, and N72-13303.

This document is presented in two volumes and is the cumulative issue of the RESEARCH DATA FACILITY (REDAF) R&D
FILE. Volume 1 lists all Earth Resources Program documentary

information that is available at the NASA-MSC. Included
in volume 2 are sensor data collected during flights over
NASA test sites and from missions flown by subcontractors
supporting the Earth Resources Survey Program. An excellent source for determining the type of data (including imagery) available, to the date of publication, at MSC.
MSC has been redesignated as the National Aeronautics and
Space Administration--Johnson Space Center.

270. U.S. National Environmental Satellite Service. ENVIRONMENTAL
SATELLITE IMAGERY: KEY TO METEOROLOGICAL RECORDS DOCU-
MENTATION NO. 5.4. Washington, D.C.: 1972-- . Illus. Paper.
Available from NTIS.

This is issued monthly to describe current cloud data obtained
by NOAA's operational environmental satellites. It contains daily global satellite imagery in condensed form as a
guide to data stored in the NOAA archives, and is designed
to assist users in selecting data for research and climatological use. The series started in November 1972 and replaces
the earlier KEY TO METEOROLOGICAL RECORDS DOCU-
MENTATION NO. 5.3 series (see item no. 271) which
dealt with camera data. Data listed are from radiometers.

Information for ordering various imagery is included with
each issue of the catalog. Many of these reports are on
microfiche, available from NTIS. An excellent and well-
organized catalog.

271. _____. KEY TO METEOROLOGICAL RECORDS DOCUMENTATION
SERIES NO. 5.3. Washington, D.C.: 1961-72. Illus. Paper.
Available from NTIS.

This was established to provide guidance information to research personnel making use of climatological data. Fre-
quently users of such data have found it necessary to spend
a great deal of time establishing whether the criteria for
observing or computing elements have changed over the period of record or in what form the data are available. It is
therefore hoped that the presentation of this series may not
only conserve valuable time but may have a direct influence
on improving the accuracy of the research results.

Chapter 5

MAPS

One of the major benefits from the use of various remote-sensing data has been the ability to generate maps at unprecedented rates. Daily maps of the weather have been based on meteorological satellite data for over a decade and, although the need for individual observations has not been decreased, the satellite data have presented a broader view of the weather situation in addition to adding information to the ground observations.

Aerial photography has been a standard source of information for providing maps of the earth's surface. Because this source is so well understood and because practically every topographic map made today has some information derived from aerial photography, this source of cartographic material has not been included in this portion of the information guide. On the other hand, the use of Landsat, radars, and other remotely sensed data sets for cartographic purposes has only begun. The cartographic products rather than the methodology of achieving that product, is the subject of this small chapter. Only a few examples of those maps which are available have been cited, and these have tended to be either experimental products which are to lead to production products or, in some cases, one-of-a-kind products which have been the result of research efforts.

In general, remote-sensing data have been used extensively to update existing maps and have contributed to the greater accuracy of such revisions. Also, numerous special interpretation maps, based solely or primarily upon remotely sensed data appear in individual papers and articles published in the open literature. These sources may well indicate the production products of the near future and should be surveyed for the latest developments.

272. Moore, Patrick. THE ATLAS OF THE UNIVERSE. New York: Rand McNally and Co., 1970. 272 p. Illus. (part col.), col. charts, ports.

> An excellent atlas of the universe which includes a section (pp. 32-79) titled "Atlas of the Earth from Space." This section includes numerous photographs from Gemini and Apollo missions, complete with descriptive comments and interpretations. Some aerial photomosaics are also included.

273. National Geographic Society. PORTRAIT USA: THE FIRST COLOR PHOTOMOSAIC OF THE 48 CONTIGUOUS UNITED STATES. Washington, D.C.: July 1976. 1 col. sheet. 74 x 107 cm. Scale 1:4, 500,000.

> This photomosaic was prepared in cooperation with the National Aeronautics and Space Administration and consists of 569 separate and essentially cloudless views of portions of the United States. Unlike most Landsat derived color maps and composite products, this map renders the United States in its approximate true and natural colors. An excellent article accompanies the map in this issue.

274. Nevada. Bureau of Mines and Geology. SATELLITE PHOTOMAP OF NEVADA--1976. Originally prepared by the Soil Conservation Service of the U.S. Department of Agriculture for NASA. Reno: Nevada Bureau of Mines and Geology, Mackay School of Mines, University of Nevada, 1976. 1 sheet. 84 x 56 cm. Scale 1:1,000,000.

> This cloud-free mosaic uses Band 5 ERTS imagery collected between 23 July and 31 October 1972. The entire state of Nevada and a small portion of adjoining states are also included. The Albers Equal Area Projection is used.

275. Rao, M.S.V.; Abbott, W.V. III; and Theon, J.S. SATELLITE-DERIVED GLOBAL OCEANIC RAINFALL ATLAS (1973 AND 1974). NASA SP-410. Washington, D.C.: Scientific and Technical Information Office, National Aeronautics and Space Administration, 1976. Va. pag. Illus., figs., refs. Available from the U.S. Government Printing Office.

> The basis of the work presented in this publication is the selective response to liquid water in the atmosphere of the Electrically Scanning Microwave Radiometer (ESMR) operating at 19.35 GHz (bandwidth of 250 MHz) carried on the Nimbus 5 satellite. A series of maps were generated based on the corrected measured brightness temperature of the ESMR. All data within one (geographical) degree of the coasts and over land, were disregarded, thus yielding maps of the global oceanic rainfall rates. This is because the oceans provide a nearly uniform background for satelliteborne radiometers; consequently the observed data can be easily corrected to represent atmospheric moisture content. This publication is an excellent example of uses of remote sensing to obtain a global data set from very isolated areas and on a continuing basis. Maps covering weeks, months, seasons, and entire years are presented in the several appendixes. Data are presented for five degree of longitude by four degree of latitude grid systems. No data poleward of 72°N latitude are available.

276. Thrower, Norman J.W. "Land Use in the Southwestern United States—from Gemini and Apollo Imagery." Map supplement in ANNALS, ASSOCIATION OF AMERICAN GEOGRAPHERS 60, no. 1 (March 1970): 208-9. Washington, D.C.: Association of American Geographers, 1970. 1 col. sheet. 91 x 98 cm. Scale 1:1,000,000.

> This map was based on Gemini 3, 4, 5 and Apollo 6 and 9 photography obtained between March 1965 and March 1969. Nine major land uses are identified and mapped: (1) transportation, (2) settlements, (3) cropland, (4) unimproved grazing land, (5) unproductive land, (6) extractive, (7) arboreal associations, (8) water bodies, and (9) some "uninterpretable areas."

277. U.S. Geological Survey. APOLLO 6 PHOTOMAPS OF THE WEST-EAST CORRIDOR FROM THE PACIFIC OCEAN TO NORTHERN LOUISIANA. Prepared in cooperation with NASA. Washington, D.C.: 1969. 4 sheets. 48 x 129 cm. Scale 1:500,000. Prints and transparencies used to prepare the APOLLO 6 PHOTOMAPS are available from Technology Applications Center, University of New Mexico, Albuquerque, N.Mex. 87106.

> This set of maps has been produced by the USGS to assess the usefulness of space photography for national resources mapping studies. The index diagram shows the photograph numbers of the Apollo vertical photography keyed to USAF ONC. Reliability diagram as well as location and coverage diagrams are included.

278. _____. FOLIO OF LAND USE IN THE WASHINGTON, D.C. URBAN AREA. USGA Miscellaneous Investigations Series Map I-858. Prepared in cooperation with NASA. Reston, Va.: 1974. 3 sheets. Scale 1:100,000.

> These maps are prototype products of experiments in land use change detection using remote-sensing aboard aircraft and earth-orbiting satellites. Contemporaneous sensor data and census data are compared for a sample of urban test sites. These maps are test demonstrations of a system of new tools used to assess and monitor urban and regional environments, especially those undergoing rapid change. The six maps are as follows: (a) I-858-A "Land Use Map, 1970, Washington Urban Area," (b) I-858-B "Annotated Orthophoto Map, 1970, Washington Urban Area," and (c) I-858-C "Census Tracts, 1970, Washington Urban Area," (d) I-858-D "Land Use Change, 1970-72," (e) I-858-E "Land Cover from Landsat, 1973, with Place Names," and (f) I-858-F "Land Cover from Landsat, 1973, with Census Tracts." All are available from the U.S.G.S. Branch of Distribution, 1200 South Eads Street, Arlington, Va. 22202.

279.   _____. HARTFORD, CONN., N.Y., N.J., MASS. (NK 18-9).
Reston, Va.: 1973. 1 sheet. 50 x 80 cm. Scale 1:250,000.

> This map is an experimental printing of a map based on
> Skylab satellite (Mission 3-S190A) photography. It covers
> the area of the USGS topographic map of the same name
> and scale. Printed in "false color" from photographs taken
> in September 1973. Some annotations are included along
> with UTM tick marks.

280.   _____. LANDSAT IMAGE MAPS. Reston, Va.: 1973-76. 14
sheets. Size varies. Scale varies. All are printed versions (not
photo products) and are available from the USGS Branch of Distribution, Arlington, Va. 22202.

> A series of photomaps, in black and white, blue tone, and
> color, these have been prepared by the USGS.

281.   _____. PHOENIX (NI 12-7) EXPERIMENTAL 1:250,000 SCALE
SPACE PHOTOMAPS. Washington, D.C.: March 1970. 3 sheets.
86 x 56 cm. each. Scale 1:250,000.

> This experimental map set holds considerable promise of im-
> proving the content, currency, and accuracy of the conven-
> tional line map by combining these maps with space derived
> imagery. The space photography used in this experiment was
> collected by a handheld camera (NASA experiment SO65)
> on Apollo 9. Two combined conventional maps and space
> photography data maps are presented and illustrate the dif-
> ference between retaining and omitting the contours of the
> conventional line map.

282. U.S. Geological Survey. Office of Chief Geographer. ATLAS OF
URBAN AND REGIONAL CHANGE--SAN FRANCISCO BAY REGION.
USGS Open File Report. Reston, Va.: U.S. Geological Survey,
1973. 44 sheets. 76 x 56 cm. Scale 1:62,500.

> These forty-four maps are the first sheets in the prototype
> looseleaf ATLAS OF URBAN AND REGIONAL CHANGE.
> They are but one part of an ongoing experiment in urban
> land use change detection using remote sensors aboard air-
> craft and satellites. Sensor data and census data were com-
> pared for a sample of the urban test sites. The efforts are
> part of the USGS, EROS program, and the NASA Earth Ob-
> servation Program.

283. U.S. Soil Conservation Service. MOSAICS OF ERTS-1 IMAGERY OF
CONTERMINOUS UNITED STATES AND ALASKA. Prepared in coop-
eration with NASA. Washington, D.C.: August 1976. Size varies.
Scale varies. Available from Cartographic Division, SCS, Federal
Building, Hyattsville, Md. 20782.

A complete photomap of the forty-eight contiguous United States using bands 5 and 7 of ERTS-1 imagery was prepared by USDA-SCS for NASA, it is constructed at a scale of 1:1,000,000 and reproduced (generally on medium weight semimatte photo paper) at scales of 1:500,000 to 1:10,000,000, which divide the United States into six, seventeen, and fifty-four individual photomap sheets, respectively. For Alaska, only band 7 was used to construct and photomap which was produced in cooperation with the Resource Planning Team of the Joint Federal State Land Use Commission for Alaska.

Chapter 6

BIBLIOGRAPHIES

Within remote sensing there are rapid changes occurring within the literature. One method for keeping up to date is to search the bibliographies. Several excellent examples (e.g., STAR) are even prepared on a bimonthly basis. Many bibliographies are prepared without annotations and are, thus, of rather limited utility--others are prepared from a very limited (or very broad) point of view and may over or under represent the area of remote sensing in which a particular researcher is interested.

This section of the remote-sensing information guide has attempted to include the many references which are available. It is, possibly, one of the most important parts of the entire guide because through the use of these references one can quickly locate the multitude of publications which are available for study. Each reference is appended with a short annotation to indicate the nature of the bibliographic materials cited. Through the use of these references, one gains access to the numerous journal and symposia articles as well as other open literature publications. The reader should note that it is for this reason that such "single subject" references have not been included in the other portions of this guide.

283A Baker, D.R.; Flanders, Allen F.; and Fleming, M. ANNOTATED BIBLIOGRAPHY OF REPORTS, STUDIES AND INVESTIGATIONS RELATING TO SATELLITE HYDROLOGY. ESSA Technical Memorandum NESCTM 10. Washington, D.C.: U.S. Department of Commerce, Environmental Science Services Administration, July 1970. ii, 28 p. Available from NTIS, N70-38529, N71-14697, and PB-194-072.

> This bibliography, consisting of approximately 110 citations, is organized by year and covers the decade of 1960 to 1970. Citations cover individual journal articles, numerous government reports, proceedings of symposia and meetings and/government hearings amongst others. The subjects covered in the citations span the range of hydrology including sea ice identification, snow mapping, cloud mapping, and precipitation. It is especially interesting to note the trend in such work during the period covered by this bibliography--much

effort was oriented toward basic instrument design, cloud studies and spectral measurements.

284. Berlin, G. Lennis. URBAN APPLICATIONS OF REMOTE SENSING. Exchange Bibliography no. 352. A Revision of CPL Exchange Bibliography no. 222. Monticello, Ill.: Council of Planning Librarians, December 1972. 65 p. Paper.

Urban oriented remote-sensing research continues at an accelerated pace, and as a consequence more and more planning groups at all levels are using remotely sensed derivative data in their programs. Emanating from both planning and research groups is a wealth of new material that has been published during the past year. For this reason Exchange Bibliography No. 222 has been updated to include studies that were completed prior to September 1972. This edition contains 691 entries, 249 more than the 1970 compilation.

285. Bidwell, Timothy C., and Mitchell, Cheryl A. AUTHOR INDEX TO PUBLISHED ERTS-1 REPORTS. Sioux Falls, S.Dak.: Technicolor Graphics, 1975. 86 p. Paper. Available from NTIS, PB-248-294, or N76-23673.

This report consists of a bibliography of papers resulting from ERTS-1 experiments between April 1972 and September 1975. A total of 334 scientific experiments are listed by author. Each listing includes author's name, address, title of experiment, NASA accession number, and NTIS order number, date of publication, number of pages in the report, and the type of report (monthly or bimonthly, comprehensive scientific and technical reports, or final report).

286. Bryan, M. Leonard. RADAR REMOTE SENSING FOR GEOSCIENCES: AN ANNOTATED AND TUTORIAL BIBLIOGRAPHY. Ann Arbor: Environmental Research Institute of Michigan, 1973. 298 p. Illus., indexes. Paper.

Consists of 383 references, with annotations and author's abstracts (where available) of papers in the open literature which deal specifically with radar remote sensing. Chapters are arranged by earth-resources subjects and two indexes, author and NTIS catalog number, are provided.

287. Carter, William D. ANNOTATED BIBLIOGRAPHY OF USGS TECHNICAL LETTERS--NASA PAPERS ON REMOTE SENSING INVESTIGATIONS THROUGH JUNE 1967. U.S. Geological Survey Open File Report. NASA Technical Letter 86. Washington, D.C.: U.S. Geological Survey, 1969. ii, 46 p.

This annotated bibliography includes reports written as a result of the feasibility studies in remote sensing by the USGS

working in cooperation with and funded by NASA under its
Earth Resources Survey Program. These reports cover a span
of three years, from inception of the program in late 1964
through June 1967. It is intended that the bibliography
will be periodically updated. The work is organized into
several major sections defined by the wavelengths of the EM
spectrum used: (1) photography, (2) radar, (3) infrared, and
(4) ultraviolet. Two additional sections are applications of
remote sensing and a list of all NASA technical letters prepared by the USGS. Although the papers are considered as
preliminary, and therefore not to be freely quoted, they are
of interest to the scholar who wishes to observe the stages
of progress of remote-sensing studies.

288. Chardon, R.E., and Schwertz, E.I., Jr. AN ANNOTATED BIBLIOG-
RAPHY OF REMOTE SENSING APPLIED TO URBAN AREAS 1950-1971.
Miscellaneous Publication no. 72. Baton Rouge: School of Geosci-
ence, Louisiana State University, 1972. vii, 39 p. Paper.

The authors have compiled a bibliography of 572 articles,
each with a short annotation under four major headings: (1)
list of remote-sensing bibliographies, (2) list of periodicals
with frequent remote-sensing articles, (3) list of general
works, and (4) index. The greatest strength of this excellent work is the index which greatly aids in locating material. However, often the annotations are too short to yield
sufficient aid in discriminating between articles.

289. Dellwig, Louis F., et al. USE OF RADAR IMAGES IN TERRAIN
ANALYSIS: AN ANNOTATED BIBLIOGRAPHY. Under contract DAAK
02-75-L-0145 to U.S. Army Engineer Topographic Laboratories, Ft.
Belvoir, Va. Lawrence: University of Kansas, August 1975. 336 p.
Paper. Available from NTIS, N76-29693.

This annotated bibliography, consists of articles, papers, and
reports dealing with the application of imaging radar systems
to the geosciences and contains approximately three hundred
entries. A short introductory statement about radar imaging
and interpretation and two indexes (author and subject) are
included.

290. Ekimov, Roza. REMOTE SENSING: LITERATURE SEARCH. Los Angeles: Humble Oil and Refining Co., March 1971. 24 p. Paper.

The objective of this literature search is to present a selective bibliography of publications introducing the field of remote sensing and its geographical applications, such as land
use analysis in agriculture, forestry, and urban and rural
planning with special emphasis in geology. A total of thirty-
one references are cited.

Bibliographies

291. Glasby, J.P., and Lowe, D.G. REMOTE SENSING BIBLIOGRAPHY
FOR EARTH RESOURCES, 1969. Washington, D.C.: Water Resources
Division, U.S. Geological Survey, May 1971. 182 p. Paper.
Available from NTIS, PB-202-726, or N71-35479.

> A compilation of 612 reference citations from 1969 litera-
> ture is given in this bibliography, the fourth in a series of
> bibliographies on remote sensing. Technical material con-
> cerning methodology and applications for earth-resources
> studies is cited, stressing advancements in airborne and earth
> satellite techniques. Nine subject areas are covered and
> each item cited includes title, bibliographic citation, and a
> set of indexing descriptors. Indexes cover citation numbers,
> author, corporate author, and subject. Seventy journals
> were searched in the preparation of this bibliography.

292. Haack, Barry N. REMOTE SENSING AND HIGHWAY TRANSPOR-
TATION PLANNING: AN ANNOTATED BIBLIOGRAPHY. Ann Arbor:
Environmental Research Institute of Michigan, February 1975. 71 p.
Index.

> This annotated bibliography contains ninty-four references
> related to remote sensing and highway planning. Its intent
> is to provide a format for individuals to become acquainted
> with the range of possible applications of remote sensing in
> this field. Entries are divided into sections on regional
> transportation planning, corridor selection and design of
> highways, materials pertinent to highway engineering, en-
> vironmental impact assessment of highways, and the economic
> and social impact of highways. Sections on additional source
> materials and an author index are included.

293. Honea, Robert B., and Prentice, Virginia L. SELECTED BIBLIOGRA-
PHY OF REMOTE SENSING. Interagency Report NASA-129, NASA-
CR-101458. Washington, D.C.: U.S. Geological Survey; Houston:
Manned Spacecraft Center, September 1968. 35 p. Paper. Avail-
able from NTIS, N68-36402.

> The objective of this monograph is to present a selective
> bibliography of publications pertaining to geographic appli-
> cations of remote-sensing techniques with special reference
> to land use classifications and analysis. The bibliography
> is divided into four parts: (1) introduction to remote sensing
> and geographic applications, (2) technical publications, es-
> sential references for research, (3) land use investigations
> using remote-sensing techniques, and (4) data handling and
> automatic data processing techniques. An 11-page supple-
> ment includes 141 references.

294. Howard, William A. REMOTE SENSING OF THE URBAN ENVIRON-
MENT: A SELECTED BIBLIOGRAPHY. Exchange Bibliography no. 69.

Montecello, Ill.: Council of Planning Librarians, March 1969. 6 p.
Paper.

Remote sensing of the urban environment potentially offers
very great possibilities for acquiring data relevant to the
planning function. To date, however, very little research
has been carried out to determine the extent to which these
techniques might serve as a data imput for planning. The
bibliography was compiled in connection with research on
developing remote sensing display modes to satisfy urban
planning data input needs. Perhaps other researchers will
find the sources cited in this bibliography of some usefulness in their work. A total of fifty-nine references are
cited.

294A   Hundemann, Audrey S. REMOTE SENSING APPLIED TO ENVIRON-
MENTAL POLLUTION DETECTION AND MANAGEMENT: A BIBLIOG-
RAPHY WITH ABSTRACTS. Progress Report, 1964–July 1978. Springfield, Va.: National Technical Information Service, August 1978.
163 p. Available from NTIS, N79-10505.

Application of remote sensing methods to air, water, and
noise pollution problems is discussed. Topic areas cover
characteristics of dispersion and diffusion by which pollutants are transported, eutrophication of lakes, thermal dis-
charges from electric power plants, outfalls from industrial
plants, atmospheric aerosols under various meteorological
conditions, monitoring of oil spills, and application of remote sensing to estuarian problems. This updated bibliography contains 156 abstracts, 23 of which are new entries to
the previous edition. (Above comment from STAR, V. 17,
No. 1, 08JAN79, p. 68).

294B   _____. REMOTE SENSING APPLIED TO GEOLOGY AND MINER-
ALOGY. A BIBLIOGRAPHY WITH ABSTRACTS. Progress Report,
1973–July 1978. Springfield, Va.: National Technical Information
Service, August 1978. 154 p. Available from NTIS, N79-10507.

The use of Landsat satellites and other remote–sensing methods in geological and mineralogical applications is discussed.
Abstracts cover rock and soil mapping, terrain analysis, direct and indirect mineral exploration, fault tectonics, and
general geologic studies of various countries. A few abstracts pertain to equipment and techniques used in the
studies. This updated bibliography contains 147 abstracts,
25 of which are new entries to the previous edition. (Above
comment from STAR, V. 17, No. 1, 08JAN79, p. 68).

294C   _____. REMOTE SENSING APPLIED TO URBAN AND REGIONAL
PLANNING: A BIBLIOGRAPHY WITH ABSTRACTS. Progress Report,
1964–July 1978. Springfield, Va.: National Technical Information

Service, August 1978. 70 p. Available from NTIS, N79-10506.

Urban and regional planning using aerial photography and satellite remote-sensing methods is discussed. Abstracts cover the use of remote sensing in land use mapping, traffic surveys and urban transportation planning, and taking inventories of natural resources for urban planning. Abstracts dealing with land use and residential quality associated with acting as an influence on health and physical well-being are included. This updated bibliography contains sixty-three abstracts, three of which are new entries to the previous edition.

295. Ivey, Nancy B., comp. REMOTE SENSING LABORATORY PUBLICATION LIST, 1964-1976. Technical Report RSL-100. Lawrence: Remote Sensing Laboratory, University of Kansas Center for Research, May 1977. 294 p.

The Remote Sensing Laboratory at the University of Kansas Center for Research was established in 1964 and presently has a staff of approximately sixty-five people. Although most of the staff are electrical engineers, the laboratory continues to maintain an interdisciplinary and multidisciplinary approach to the field of remote sensing. The contents of this volume reflect the diversity of backgrounds of the laboratory staff. This publication list provides a reference to all journal articles and technical reports published by members of the Remote Sensing Laboratory during the thirteen-year period of 1964-76 and is intended to facilitate the acquisition of these documents by interested members of the scientific community. The bibliography is divided into two major sections containing (1) "Technical Reports and Publications" and (2) "Technical Memoranda." Approximately four hundred and fifty "Technical Reports and Publications" (including Ph.D. dissertations) and two hundred and fifty "Technical Memoranda" are included. Ordering information is given and it is noted that this publication list will be updated annually.

296. Jones, Natalie E. BIBLIOGRAPHY OF REMOTE SENSING OF RESOURCES. Prepared for the Earth Resources Survey Program, Space Applications Program, NASA. Ft. Belvoir, Va.: U.S. Army Corps of Engineers, September 1968. 40 p. Available from NTIS, N68-19870.

This bibliography is a compilation of references found in the open literature concerning remote sensing of resources and associated subjects. Generally, the references are drawn from publications between January 1960 and June 1966. No references on the technical aspects of electromagnetic sensing, geophysics, or meteorology have been included.

The bulk of the references have been drawn from AVIATION WEEK AND SPACE TECHNOLOGY, PHOTOGRAMMETRIC ENGINEERING, and the University of Michigan Remote Sensing Symposia.

297.   Kracht, James B., and Howard, William A. APPLICATIONS OF RE-MOTE SENSING, AERIAL PHOTOGRAPHY AND INSTRUMENTED IMAGERY INTERPRETATION TO URBAN AREA STUDIES. Exchange Bibliography no. 116. Monticello, Ill.: Council of Planning Librarians, December 1970. 34 p. Paper.

This bibliography was developed in connection with contin-uing research concerning the development of appropriate re-mote-sensing display modes to satisfy the information require-ments of urban and urban region information systems. Sec-tions of the bibliography consist of the following: (1) basic introductory sources, (2) sources relating to urban and re-gional analysis, (3) sources relating to highway and traffic studies, (4) sources relating to instrumented systems for in-terpreting, compiling, and mapping, and (5) bibliographies. Approximately 425 items are cited.

298.   Krumpe, Paul F. REMOTE SENSING OF TERRESTRIAL VEGETATION: A COMPREHENSIVE BIBLIOGRAPHY. Knoxville: Graduate Program in Ecology, University of Tennessee, January 1972. 69 p.

This bibliography contains more than eight hundred and fifty references dealing with the utilization and application of remote sensing in forestry, agriculture, plant ecology, and several closely allied fields. In addition, technical back-ground, historical, and data manipulation and analysis ref-erences are included. References are by author, publica-tion date, and title. The emphasis has been placed on ma-terials published since 1955.

299.   _____. THE WORLD REMOTE SENSING BIBLIOGRAPHIC INDEX. Fairfax, Va.: Tensor Industries, 1976. ix. 619 p.

This volume is a geographic index bibliography of over four thousand references on remote sensing of natural and agri-cultural resources throughout the world. Citations from 1970 to August 1976 are arranged within fourteen major dis-ciplines among more than one hundred and fifty geographic areas, states, and countries. This extensive compilation originates from more than eight hundred and fifty foreign and domestic sources among six major publishing categories. Instructions for procuring desired publications or reports are included in addition to guidelines for efficient utilization of the document. This book is designed as an integrated reference guide for use in remote sensing and environmental education, training, applications research, analysis, and technology. No abstracts are included with the citations.

Bibliographies

300. Llaverias, Rita K. BIBLIOGRAPHY OF REMOTE SENSING OF EARTH RESOURCES FOR HYDROLOGICAL APPLICATIONS, 1960-1967. NASA Technical Letter-134, NASA-TM-61717. Washington, D.C.: Water Resources Division, U.S. Geological Survey, November 1968. 75 p. Paper. Available from NTIS, N69-28505.

> This preliminary bibliography was prepared to acquaint hydrologists with the basic literature involved in this field. Some of the references concern specific hydrologic topics or specific remote-sensing methods. Other references on vegetation mapping and geology were included so that the reader can find information on the selection, processing, and use of remote-sensing data in these cognate fields. References are listed in alphabetical order by author. This is the first in a series of USGS bibliographies dealing with remote sensing and is rather difficult to use because it lacks both indexes and abstracts. It is a working bibliography and should be considered as such. All phases of remote sensing are covered in approximately four hundred and fifty citations.

301. _____. REMOTE SENSING BIBLIOGRAPHY FOR EARTH RESOURCES, 1966-67. NASA Interagency Report-189. Washington, D.C.: Water Resources Division, U.S. Geological Survey, May 1970. 136 p. Indexes. Paper. Available from NTIS, PB-192-863.

> This is the second of a series of bibliographies on applications of remote sensing to earth resources. Earth resources in general, with the emphasis on hydrology, comprise the subject for literature citations covering 1966-67. Included are 412 citations indexed by author, corporate source, and subject. One of the strong points of this bibliography is the completeness of the indexes. No abstracts are given.

302. Llaverias, Rita K., and Lowe, D.G. REMOTE SENSING BIBLIOGRAPHY FOR EARTH RESOURCES, 1968. NASA Interagency Report-203. Washington, D.C.: Water Resources Division, U.S. Geological Survey, October 1970. xii, 246 p. Paper. Available from NTIS, PB-195-748.

> Citations covering the field of earth resources, sensors, and sensor techniques are included. Personal and corporate author indexes are provided for most of the 801 citations. Each subject index contains two or more descriptions for each item. None of the papers included in this bibliography is abstracted.

303. McGinnies, W.G. AN ANNOTATED BIBLIOGRAPHY AND EVALUATION OF REMOTE SENSING PUBLICATIONS RELATING TO MILITARY GEOGRAPHY OF ARID LANDS. Technical Report 71-27-ES, Series ES-61. Natick, Mass.: Earth Sciences Laboratory, U.S. Army

Natick Laboratories, September 1970. 107 p. Available from NTIS, AD-723-061.

A comprehensive review has been made of remote-sensing publications relating to military geography of arid lands. These have been abstracted or annotated and arranged in tables relating devices and processes to geographic features. The devices and processes include black and white, color, and infrared photographs and devices utilizing longer wavelengths such as radar. Vehicles include conventional airplanes and spacecraft. Each of the 355 references is rated as being especially useful, useful, or of little value.

304. Manji, Ashraf S. USES OF CONVENTIONAL AERIAL PHOTOGRA-PHY IN URBAN AREAS: REVIEW AND BIBLIOGRAPHY. NASA Technical Letter-131. Washington, D.C.: U.S. Geological Survey, September 1968. 39 p.

This paper is divided into two parts. Part 1 presents a review of some of the past uses of aerial photography in urban areas. The empirical studies reviewed are divided into two broad categories depending on whether inventory compilation is by direct or indirect observation. Studies using inventory by direct observation are further divided into land use studies and transportation studies. Part 2 identifies some problem areas for future research. A total of 104 items are cited in the bibliography.

305. New Mexico, University of. Technology Application Center. QUAR-TERLY LITERATURE REVIEW OF THE REMOTE SENSING OF NATURAL RESOURCES. Albuquerque, N.Mex.: 1974-- .

This abstract journal (see item no. 353), organized by categories with keyword and author indexes, includes pertinent literature from INTERNATIONAL AEROSPACE ABSTRACTS (IAA) and SCIENTIFIC AND TECHNICAL REPORTS (STAR) by NASA, the ENGINEERING INDEX MONTHLY, SELECTED WATER RESOURCES ABSTRACTS, GOVERNMENT REPORTS ANNOUNCEMENTS, BIBLIOGRAPHY AND INDEX OF GEOLOGY, and NUCLEAR SCIENCE ABSTRACTS. Annual subscriptions are available. This bibliography supersedes REMOTE SENSING (1962-67) and annual updates for 1968 through 1973, previously compiled by the Technology Application Center from most of the same sources. Copies of older publications and prices are available upon request from TAC.

306. Nunnally, Nelson R. BIBLIOGRAPHY OF REMOTE SENSING APPLICATIONS FOR PLANNING AND ADMINISTRATIVE STUDIES. U.S. Geological Survey Interagency Report-234. Washington, D.C.: U.S. Geological Survey, December 1971. 65 p. Paper. Available from NTIS, N72-24423.

This bibliography, prepared under a project funded by the
Tennessee Valley Authority, has three major aims: (1) to
include representative materials, (2) avoid redundancy, and
(3) choose material of most interest to the TVA. The materials are arranged alphabetically by author, and a subject
listing covering most of the topics is included to aid in use
of the bibliography. All items are annotated--generally
with an author's abstract and the compiler's comments. A
total of 162 items are included.

307. Pearson, Bradley R. REMOTE SENSING AND URBAN ANALYSIS: A
PROJECT SUPPORTIVE BIBLIOGRAPHY. U.S. Geological Survey.
Interagency Report no. 204. Washington, D.C.: U.S. Geological
Survey, September 1970. 19 p. Paper. Available from NTIS, PB-
199-123.

The bibliography, with a heavy emphasis on aerial photography, is organized solely along the lines of an author index. It is a literature survey of directly pertinent (eighty-
seven) and indirectly pertinent (eighty-four) citations dealing
with urban applications of remote sensing.

308. Rajarajeswari, I., comp. BIBLIOGRAPHY ON REMOTE SENSING.
Trivandrum, India: Vikram Sarabhai Space Center, Indian Space Re-
search Organization, March 1977. 246 p. Indexes. Available from
NTIS, N77-21531.

The bibliography was compiled with a view to bring forth
all reports, patents, and journal articles on remote sensing
which were indexed in STAR and IAA during 1970-75. Entries are arranged alphabetically by author or title covering
details such as report number, source, and so forth. An
alphabetical key word index and a report number index are
provided, thus, aiding in the use of this reference material.

308A Sers, Sidney W. REMOTE SENSING IN HYDROLOGY: A SURVEY
OF APPLICATIONS WITH SELECTED BIBLIOGRAPHY AND ABSTRACTS.
Remote Sensing Center Technical Report RSC-22. College Station:
Texas A & M University, October 1971. i, 26 p. Index. Available
from NTIS, N72-21357.

The purpose of this document is to demonstrate remote infrared sensing as a water exploration technique. Applications described include location of aquifers, geothermal water, water trapped by faults, springs and water in desert
areas. A short introduction and text which gives a nice
summary of the state-of-the-art introduces the paper. Approximately forty-five articles are abstracted.

309. Stafford, Donald B., ed. AERIAL REMOTE SENSING: A BIBLIOG-
RAPHY. Bibliography Series, WRSIC 73-211. Washington, D.C.:

Water Resources Scientific Information Center, May 1973. 488 p.
Indexes. Available from NTIS, PB-200 163, or N73-30383.

This report containing 272 abstracts, is another in a series
of planned bibliographies in water resources to be produced
from the information base comprising SELECTED WATER RE-
SOURCES ABSTRACTS (SWRA). At the time of the search
for this bibliography, the data base had approximately
50,600 abstracts covering SWRA through 15 December 1972
(vol. 5, no. 24). The arrangement of the bibliography is
to include computer listings of significant descriptions, the
bibliography proper with abstracts and complete citations,
a comprehensive index, and an author index.

310. Stafford, Donald B.; Bruno, Richard O.; and Goldstein, Harris M.
AN ANNOTATED BIBLIOGRAPHY OF AERIAL REMOTE SENSING IN
COASTAL ENGINEERING. U.S. Coastal Engineering Research Center.
Miscellaneous Paper no. 2-73. Fort Belvoir, Va.: Coastal Engineer-
ing Research Center, U.S. Army Corps of Engineers, 1973. 122 p.
Indexes. Available from NTIS, AD-766720, and N74-12202.

This is a bibliography of representative literature covering
the applications of aerial remote-sensing techniques to coastal
engineering. About two thousand references published since
1934 are presented. Annotations accompany each bibliographic entry and are a concise and informative summary of
the references describing the characteristics of each remote
sensor in coastal engineering investigations. Computer indexes of authors, titles, and keywords are included.

311. Steiner, Dieter. "Annotated Bibliography of Bibliographies on Photo
Interpretation and Remote Sensing." PHOTOGRAMMETRIA 26, no. 4
(October 1970): 143-61.

This bibliography of bibliographies, an outgrowth of a paper
presented at the International Congress of Photogrammetry
(1968, Lausanne, Switzerland), references 254 articles. The
author estimates that these give access to approximately
50,000 additional references. Ongoing bibliographies and documentation are not included, but fourteen such information
sources are presented for the reader's convenience. Many
references to European and Russian literature are included,
together with numerous government publications. Some items
do not have annotations, or they are so brief as to be of
little use. A keyword index of approximately one hundred
items is included.

312. Thompson, William I., III. EARTH SURVEY BIBLIOGRAPHY: A KWIC
INDEX OF REMOTE SENSING INFORMATION. Prepared for National
Aeronautics and Space Administration. Report no. DOT-TSC-NASA-
70-I. Cambridge, Mass.: Transportation System Center, February

1971. 265 p. Indexes. Available from NTIS, N71-26398 and N70-42766.

This bibliography represents a collection of 1,650 bibliographic citations on remote sensing of the physical characteristics of the earth, and is intended to be used as a source document leading to additional information. A short introductory statement gives instructions on using the bibliography. A list of source bibliographies, a short list of information centers specializing in remote-sensing information, and indexes by author and subject are included. No abstracts or resumes of the articles are contained in this work.

312A   Todd, William J. A SELECTIVE BIBLIOGRAPHY: REMOTE SENSING APPLICATIONS IN LAND USE AND LAND COVER TASKS. Sioux Falls, S.Dak.: Technicolor Graphic Services, April 1978. ii, 33 p. Available from NTIS, PB-283-027 or N78-10509.

Over three hundred citations, primarily from 1968 to 1977, are listed in this bibliography dealing with applications of remote-sensing techniques to regional and metropolitan land use and land cover identity and analysis. There are no cross references, keywords or similar indices, and the citations are listed by senior author name. Thus this bibliography is relatively difficult to use. Citations only are included, summaries and annotations are lacking.

313.   U.S. National Aeronautics and Space Administration. Scientific and Technical Information Office. EARTH RESOURCES: A CONTINUING BIBLIOGRAPHY WITH INDEXES. NASA SP-7041. Washington, D.C.: U.S. Government Printing Office, 1974-- . Irregular, four to five times per year. Available from NTIS.

This publication (see item no. 366) is a continuation of NASA SP-7036 (item no. 314). It consists of a selection of annotated references to unclassified reports and journal articles that were introduced to the NASA Scientific and Technical Information System. This publication is a continuing bibliography, combining many of the STAR and IAA reports, but having a slightly different set of categories. It includes excellent documentation, and a series of five indexes: (1) subject, (2) personal author, (3) corporate source, (4) contract number, and (5) accession number. An extremely valuable bibliography and resource tool of which a total of nineteen issues (to October 1976) averaging approximately six hundred citations, have been issued. See also citation no. 331.

314.   _____ . REMOTE SENSING OF EARTH RESOURCES: A LITERATURE SURVEY WITH INDEXES. NASA SP-7036. Washington, D.C.: U.S. Government Printing Office, 1970. ix, 1221 p. Paper.

This literature survey lists 3,684 reports, articles, and other documents introduced into the NASA scientific and technical information system between January 1962 and February 1970. Emphasis is placed on the use of remote sensing and geophysical instrumentation in spacecraft and aircraft to survey and inventory natural resources and urban areas. Subject matter is grouped according to agricultural and forestry, environmental changes and cultural resources, geodesy and cartography, geology and mineral resources, oceanography and marine resources, hydrology and water management, data processing and distribution systems, instrumentation and sensors, and economic analysis. Abstracts, subject, author, and corporate source indexes are included together with indexes by contract numbers. Supplemented by NASA SP-7036 (01) (item no. 315) and continued as NASA SP-7041 (item no. 313).

315. _____. REMOTE SENSING OF EARTH RESOURCES: A LITERATURE SURVEY WITH INDEXES, 1970-1973 SUPPLEMENT. 2 vols. NASA SP-7036(01). Washington, D.C.: U.S. Government Printing Office, February 1975. 654 p. Paper.

This bibliography is a continuation of NASA SP-7036 (item no. 314) and lists 4,930 reports, articles, and other documents introduced into the NASA scientific and technical information system between March 1970 and December 1973. This publication is issued in two sections: Section 1, abstracts, and section 2, indexes. Each entry in the abstract section contains a citation and an abstract. The index section contains five indexes: (1) subject, (2) author, (3) source, (4) contract number, and (5) report number. This bibliography is continued as NASA SP-7041 (see item no. 313).

315A Vogel, Theodore C., and Books, James E. A SELECTED BIBLIOGRAPHY OF CORPS OF ENGINEERS REMOTE SENSING REPORTS. Ft. Belvoir, Va.: U.S. Army Engineer Topographic Laboratories, November 1977. 226 p. Available from NTIS, N78-19584 or AD A049351.

The purpose of this bibliography is to present a selected listing of the remote sensing technology research, technical, and contract reports, including professional papers, published by the U.S. Army Corps of Engineers divisions, districts, and laboratories. The reports are divided into twenty subject categories and listed, within each category, alphabetically by author.

316. Walters, Robert L. RADAR BIBLIOGRAPHY FOR GEOSCIENTISTS. Lawrence: Center Research in Engineering Sciences, University of Kansas, March 1968. iv, 28 p. Paper. Available from NTIS, N69-30303.

Bibliographies

The purpose of this bibliography is to provide a comprehensive source of background information emphasizing geological, agricultural, geographical, and related interpretations of modern high-resolution, side-looking airborne radar imagery. Two hundred sixty-six references are cited and indexed, covering a broad spectrum of subjects from applied imagery analysis and interpretations to selected theoretical studies. The entries are listed by author, but lack annotations. It does an especially good job of covering both journal and unpublished reports which are often difficult to find. As the title indicates, the bibliography is oriented toward geoscientists--which may be interpreted to mean natural sciences and cartography. No listed articles deal primarily with urban, social, or environmental aspects of radar remote sensing, although these are valid areas in the discipline of geography.

317. Westerlund, Frank V. URBAN AND REGIONAL PLANNING UTILIZATION OF REMOTE SENSING DATA: A BIBLIOGRAPHY AND REVIEW OF PERTINENT LITERATURE. Washington, D.C.: U.S. Geological Survey EROS Program; Seattle: Dept. of Urban Planning, University of Washington, 1972. 232 p. Paper. Available from NTIS, PB-211-101.

In order to establish relationships between capabilities of the different remote sensors and the whole range of data needs peculiar to planning, this study develops an urban planning perspective of the remote-sensing field and from that perspective provides an overview of relevant aspects of applications research. This is in the form of a survey of literature in which related studies are identified by subject and collectively described, followed by separate annotations of selected individual papers.

Chapter 7

JOURNALS

Only a few journals dealing specifically with remotely sensed data, its management, interpretation, and utility are presently available. However, many disciplines dealing with earth science, technology, engineering, and the humanities have within their separate journals, published and supported the efforts of the remote-sensing community. Hence, one may find publication of papers and articles dealing with remote sensing in a vast array of sources from the daily paper to some highly academic sources. The orientation of the articles obviously reflect the audience as well as the editorial license which is employed--where some deal with a very detailed examination of the concepts and theory, others will be concerned primarily with the nature of the interpretation and application of the information. A third group is concerned primarily with the presentation of the information and leaves the details up to the reader. An attempt, through the annotations, has been made to identify the depth and breadth with which each of these journals deal with remote-sensing data and applications. Wherever possible, information concerning source publication, frequency of publication, and similar basic data has been included. For those items identified as being available from University Microfilms, contact University Microfilms International, A Xerox Company, 300 North Zeeb Road, Ann Arbor, Mich. 48106.

318. AMERICAN ASSOCIATION OF PETROLEUM GEOLOGISTS. Bulletin. American Association of Petroleum Geologists, P.O. Box 979, Tulsa, Oklahoma 74101. 1917-- . Monthly. ISSN 0002-7464.

> This publication is one of the leading geological publications in the area of petroleum geology and occasionally contains articles dealing with remote sensing in this area of research and applications. Volume 60, no. 5 (May 1976) contains an extensive article dealing with the utility of Landsat data for such work. The journal is available on microform from University Microfilms. In addition to articles, the journal contains book reviews, bibliographies, numerous illustrations, advertisements from companies and individuals in the field, and a cumulative ten-year index.

319. AMERICAN CARTOGRAPHER. American Congress on Surveying and Mapping, 430 Woodward Building, Fifteenth Street, N.W., Washington, D.C. 20005. 1974-- . Semiannual. ISSN 0094-1689.

> This journal reflects the breadth of the field of cartography, ranging from the topographic to the thematic, from the history of mapping to map collecting, from automation to the aesthetic. Hence, it is aimed at all aspects of cartography and, as such, occasionally deals with topics of concern to the boundary between cartography and remote sensing, including aerial photography. Several articles concerning the cartographic aspects of ERTS and Landsat imagery have appeared in its pages.

320. ANNALS OF THE ASSOCIATION OF AMERICAN GEOGRAPHERS. Association of American Geographers, 1710 Sixteenth Street, N.W., Washington, D.C. 20009. 1911-- . Quarterly. ISSN 0004-5608.

> This journal, covering the field of geography, occasionally has good articles dealing with the applications of remote sensing to some specific geographical problem. Articles tend to be somewhat qualitative. Several maps (see map section of this guide) derived from remotely sensed data, have been published as their "Map Supplement."

321. APPLIED OPTICS. Optical Society of America, 200 L. Street, N.W., Washington, D.C. 20036. 1962-- . Monthly. ISSN 0003-6935.

> This journal contains articles about theoretical and experimental investigations that contribute new knowledge or understanding of any optical phenomenon, principles, or methods. It is therefore of minimal direct interest to the earth-resources (application) community but is an excellent source for keeping up with advances in optics, and therefore, the new instruments and techniques which are becoming available for new and revised remote sensor applications. Text is in German, French, Russian or (predominantly) English. The journal contains charts, illustrations, a cumulative index, and book reviews.

322. ARCTIC. Arctic Institute of North America, University Library Tower, 2920 Twenty-fourth Avenue, N.W., Calgary, Alberta, Canada T2N 1N4. 1948-- . Quarterly. ISSN 0004-0843.

> This journal deals with all aspects of the Arctic (and occasionally the Antarctic) regions including terrestrial, oceanographic, biological, and social and cultural subjects. Written in English and containing abstracts in English, French, and Russian, research results dealing with remotely sensed data are published periodically. Contains bibliographies, charts, illustrations, book reviews, letters, and commentaries.

323. ASTRONAUTICS AND AERONAUTICS. American Institute of Aeronautics and Astronautics, Inc., 1290 Avenue of the Americas, New York, N.Y. 10019. 1957-- . Monthly. ISSN 0004-6213.

This journal is oriented primarily toward engineering aspects and design of flight hardware and equipment. As such, it often contains detailed reports about various satellites, including those oriented toward earth-resources remote sensing as well as their equipment and experiments. Volume 15, nos. 4-6 (1977) for example, contains a special section on the Space Shuttle payloads, giving an excellent background and understanding that will be helpful to earth remote-sensing planners using this system.

324. AVIATION WEEK AND SPACE TECHNOLOGY. McGraw-Hill, Inc., 1221 Avenue of the Americas, New York, N.Y. 10020. 1916-- . Weekly. ISSN 005-2175.

This magazine deals primarily with military hardware developments, but also includes commentary and articles concerning various earth-resources satellite operations. Occasionally imagery and interpretations (primarily of civil and military strategic interest) of satellite data are included. An excellent source for keeping abreast of the latest plans and developments, both in the political and the technological spheres, it contains advertisements and illustrations. It is available on microfilm from Xerox, University Microfilms.

325. BIBLIOGRAPHY AND INDEX OF GEOLOGY. Formerly BIBLIOGRAPHY AND INDEX OF GEOLOGY EXCLUSIVE OF NORTH AMERICA (ISSN 0006-1520). Geological Society of America, 3300 Penrose Pl., Boulder, Colo. 80301. 1969-- . Monthly. ISSN 0006-1522.

This bibliography, complete with rather extensive abstracts which are gleaned from the authors' published abstracts, is very complete and current. It has been expanded to cover the entire range of geologic literature of the world. This is an excellent source for all geology, including remote sensing as applied to geological problems.

326. BILDMESSUNG UND LUFBILDWESEN: ZEITSCHRIFT FÜR PHOTO-GRAMMETRIE UND FERNERKUNDUNG. Deutsche Gesellschaft fur Photogrammetrie, Rheinstrasse 122, Karlsruhe 21, West Germany. 1926-- . Quarterly. ISSN 0006-2421.

This is one of the better European journals dealing with aerial photography and photogrammetry. It is often difficult to find in American libraries. Summaries of articles are published in French, English, and German. The journal contains book reviews, charts, illustrations, and references, in addition to the scientific articles.

327. BRITISH INTERPLANETARY SOCIETY JOURNAL. The British Interplanetary Society, 12 Beesbourough Gardens, London, SW1V 2JJ, Engl. 1934-- . Monthly. ISSN 0007-084X.

This journal occasionally contains earth resources papers, but is concerned primarily with space applications of various navigation and communication satellites and systems. Volume 28, no. 9-10 (September-October 1975), has a large section (pp. 595-688) devoted to space applications for earth-resources projects. In this issue, ERTS is the primary sensor of concern. Other recent issues devoted to remote sensing include volume 30, no. 5 (May 1977), and volume 31, no. 1 (January 1978).

328. CANADIAN JOURNAL OF EARTH SCIENCES. National Research Council of Canada, Ottawa, K1A OR6, Canada. 1964-- . Monthly. ISSN 0008-4077.

This journal, devoted to the earth sciences, strongly favors geology, minerology, hydrology, and physical geography. Occasional articles concerning remote sensing as applied and used in these disciplines are included. Contains bibliographies, illustrations, and an index.

329. CANADIAN JOURNAL OF REMOTE SENSING. Canadian Aeronautics and Space Institute, 77 Metcalfe Street, Ottawa, Ont., Canada K1P 5LP. 1975-- . Bi-monthly. ISSN 0008-2821.

This journal, an organ of the Canadian Remote Sensing Society (a constituent society of the Canadian Aeronautics and Space Institute), publishes papers, articles, and technical notes dealing with remote-sensing applications, instrumentation, and theory. Also included is the remote sensing news of the society and notices of upcoming events.

330. CONTRIBUTIONS TO GEOLOGY. University of Wyoming, Department of Geology, P.O. Box 3006, Laramie, Wyo. 82070. 1962-- . Semiannual. ISSN 0010-7980.

This journal deals primarily with contributions to the geology of Wyoming and adjacent western states. A special issue (edited by R.B. Parker) of the winter of 1973 (volume 12, no. 2) dealt with ERTS data. Ten articles dealt specifically with scenes and geology of Wyoming and provide a wide array of approaches to both manual and machine interpretation philosophies.

331. EARTH RESOURCES: A CONTINUING BIBLIOGRAPHY WITH INDEXES. Springfield, Va.: National Technical Information Service; Washington, D.C.: U.S. Government Printing Office, 1974-- . Irregular. NASA SP-7041. U.S. National Aeronautics and Space Administration,

Scientific and Technical Information, National Technical Information Service, Springfield, Va. 22161.

See citation no. 313.

332.   ENVIRONMENTAL SCIENCE AND TECHNOLOGY. American Chemical Society, 1155 16th St., N.W., Washington, D.C. 20036. 1967-- . Monthly. ISSN 0013-936X.

This journal occasionally contains articles oriented toward remote sensing as applied to environmental problems (e.g., water pollution) and the remote-sensing programs of various organizations and government agencies. The journal is available in microform from the American Chemical Society.

333.   FUNCTIONAL PHOTOGRAPHY. Formerly PHOTOGRAPHIC APPLICATIONS IN SCIENCE AND TECHNOLOGY, and PHOTOGRAPHIC APPLICATIONS IN SCIENCE, TECHNOLOGY AND MEDICINE. PTN Publishing Corp., 250 Fulton Ave., Hempstead, N.Y. 11550. 1967-- . Six issues per year. ISSN 0031-871X.

This journal deals with all aspects of photography in the sciences, medicine, and technology (e.g., microphotography, darkroom techniques). There are occasionally articles dealing with airborne and satelliteborne applications of remote-sensing information. Recent articles have dealt with the assessment of damage following earthquakes, nuclear power plant site evaluation, and use of wetlands. The journal contains book reviews, abstracts, charts, and illustrations.

334.   GEO-ABSTRACTS-G: REMOTE SENSING AND CARTOGRAPHY. University of East Anglia, Norwich, NOR 88C, Engl. 1974-- . Bimonthly. ISSN 0305-1951.

Although many of the references included are up to two years old (since their publication), material included in this abstracting journal comes from a wide variety of sources including government reports, journal articles, and some unpublished (informal) reports. Nonwestern literature, especially from USSR and Eastern Europe are included and thus, this journal greatly expands the references available to the Western reader. A cumulative index is prepared each five years. Abstracts, often directly or abridged from the authors' abstracts, form the bulk of this publication.

335.   GEOFORUM: THE INTERNATIONAL MULTI-DISCIPLINARY JOURNAL OF PHYSICAL, HUMAN AND REGIONAL GEOSCIENCES. Pergamon Press, Inc., Maxwell House, Fairview Park, Elmsford, N.Y. 10523. 1970-- . Bimonthly. ISSN 0016-7185.

This journal deals primarily with contemporary issues and problems which are of concern to the geoscience community.

Papers are accepted in French, German, or English, with abstracts in the same three languages. Volume 2 (1970) was a special issue edited by H.Th. Verstappen and dealt with aerospace observation techniques. The journal contains book reviews, charts, and illustrations, and is available on microfilms.

336. BULLETIN OF THE GEOLOGICAL SOCIETY OF AMERICA. c/o Ed. Bennie W. Troxel, Geological Society of America, 3300 Penrose Pl., Boulder, Colo. 80301. 1888-- . Monthly. ISSN 0016-7606.

One of the leading geologically oriented journals of the world, it contains occasional articles on remote sensing of various geological features. An excellent source for ongoing and recent research and results. A cumulative index is published each decade. It also contains illustrations, abstracts, and bibliographies.

337. GEOTIMES: NEWS OF THE EARTH SCIENCES. American Geological Institute, 5205 Leesburg Pike, Falls Church, Va. 22041. 1956-- . Twelve issues per year. ISSN 0016-8556.

Basically, this journal is a newsletter and, as such, contains numerous reports of ongoing research programs, listings of upcoming meetings (for several years in advance), listings of new books, reports and maps, and numerous advertisements concerning, among others, remote-sensing activities. An excellent source for keeping abreast of activities in the field.

338. GOVERNMENT REPORTS ANNOUNCEMENTS AND INDEX. U.S. Department of Commerce, National Technical Information Service, Springfield, Va. 22151. 1961-- . Biweekly. ISSN 0097-9007.

This journal is published to simplify and improve public access to new report literature as it becomes available. The two publications (GRA, GRI) must be used together, for GRI provides the index to aid in finding abstracts in GRA. The index is by subject, corporate author, contract and grant number, personal author, and accession and report number. Most reports are available in both microfiche and hard copy. This series provides most of the same type of coverage as noted in STAR and IAA, as well as a broader disciplinary coverage. Prior to March 1961, the same type of abstract service was conducted under the name of UNITED STATES GOVERNMENT RESEARCH AND DEVELOPMENT REPORTS (USGRDR) which was essentially the same as GRA. At that time the index (i.e., GRI) was produced annually.

339. IEEE PROCEEDINGS. Institute of Electrical and Electronics Engineers,

Inc., 345 E. Forty-seventh St., New York, N.Y. 10017. 1913-- . Monthly. ISSN 0018-9219.

This journal is primarily review and tutorial papers covering all aspects of electrical engineering and research papers not covered in one of the other IEEE publications. Remote sensing is approached primarily from the engineering and technological points of view. The journal contains book reviews, abstracts, illustrations, and figures. Occasional special issues deal with items of special interest to the earth sciences community, for example, SPECIAL ISSUE ON REMOTE ENVIRONMENTAL SENSING 57, no. 4 (April 1969); ISSUE ON MODERN RADAR TECHNOLOGY AND APPLICATIONS 62, no. 6 (June 1974).

340. IEEE, TRANSACTIONS ON GEOSCIENCE ELECTRONICS. Institute of Electrical and Electronics Engineers, Inc., 345 E. Forty-seventh St., New York, N.Y. 10017. 1963-- . Quarterly. ISSN 0018-9413.

The primary concern of this journal is the publication of both theoretical and applied papers on geoscience and electronics and generally not having an outlet in existing geophysical or engineering journals. Many articles dealing with the developments of SLAR and SAR instruments and their applications have appeared in this journal. Several special issues of interest to the remote-sensing community have been published:

REMOTE SENSING 9, no. 3 (July 1971).
SUMMARY OF ERTS EXPERIMENTS 11, no. 1 (January 1973).
DATA COLLECTION FROM MULTIPLE EARTH PLATFORMS 13, no. 1 (January 1975).
MACHINE PROCESSING OF REMOTELY SENSED DATA 15, no. 3 (July 1977).

341. INTERNATIONAL AEROSPACE ABSTRACTS (IAA). Technical Information Service. American Institute of Aeronautics and Astronautics, 750 Third Ave., New York, N.Y. 10017. 1961-- . Semimonthly. ISSN 0020-5842.

This publication presents abstracts in the fields of aeronautics and space science and technology. The abstract and index service uses the same key words and subject categories as does STAR, and therefore they complement one another. Periodicals, conference papers, and journals as well as article translations are covered. The coverage of conference papers is the major contribution of this journal. Author, title, contract number, IAA accession number, and AIAA meeting papers are all included as major indexes. It contains annual cumulative indexes.

342. ITC JOURNAL. International Institute for Aerial Survey and Earth
Sciences, P.O. Box 6, Enschede, Netherlands. 1973-- . Quarterly.
No ISSN.

Although this journal is essentially an in-house publication
of ITC, with the majority of the articles being prepared by
the staff and students of that institution, it is of major importance to the earth-resources remote-sensing community.
New ideas, reports about ongoing research, and the development and application of new remote-sensing methodologies
comprise the bulk of the publication. Strong points deal
with the use of aerial photography and associated equipment
and methods, the development of cartographic materials and
the applications of IR and radar to earth-resources problems.
The journal contains book reviews, charts, illustrations, tables, references, advertisements, and in-house notices of
the developments at the institute.

343. JOURNAL OF ASTRONAUTICAL SCIENCES. American Astronautical
Society, 6060 Duke St., Alexandria, Va. 22304. 1954-- . Quarterly. ISSN 0021-9142.

The American Astronautical Society is a national organization dedicated to the advancement of the astronautical sciences. The society considers manned interplanetary space
flight a logical progression from today's earth satellite operations although not primarily oriented to earth science disciplines. Occasional articles dealing with earth orbiting
satellite navigation, ground tracks, and observations are published in this journal. See for example, vol. 22, no. 2
and 3 (April-June 1975) and vol. 22, no. 3 (July-Septem-
ber 1975), for articles on Skylab data applications.

344. JOURNAL OF ENVIRONMENTAL SCIENCES. Institute of Environmental Sciences, 940 E. Northwest Hwy., Mt. Prospect, Ill. 60056.
1958-- . Bimonthly. ISSN 0022-0906.

"The Institute of Environmental Sciences is a professional
society of engineers, scientists, and educators simulating
and testing in the environments of earth and space, for the
betterment of mankind and the advancement of industry and
science." As such, their journal covers a wide range of
subjects, topics, and approaches including articles on remote sensing of the earth. Vol. 17, no. 2 (March/April
1974), was devoted to ERTS related studies.

345. JOURNAL OF FORESTRY. Society of American Foresters, 1010 Sixteenth St., N.W., Washington, D.C. 20036. 1902-- . Monthly.
ISSN 0022-1201.

This leading journal in the field of forestry reports on the science, practice, and profession of forestry. Remote-sensing articles, especially dealing with the use of methods for detecting stress (moisture, disease) on vegetation and timber volume studies are included. Use of aerial photography and infrared for these purposes are predominant. It also contains book reviews, abstracts, illustrations; a cumulative index is published each decade.

346. JOURNAL OF GEOPHYSICAL RESEARCH. American Geophysical Union, 1909 K St., N.W., Washington, D.C. 20006. 1896-- . Three issues per month. ISSN 0022-1406.

A rather scholarly journal which deals primarily with the more theoretical and speculative applications of remote-sensing data and interpretation techniques, along with other topics of interest to the geophysical community. An excellent way to keep abreast of the "cutting edge" of present research in the areas of remote-sensing studies and applications. It contains abstracts, illustrations, and bibliographies.

347. JOURNAL OF GLACIOLOGY. International Glaciological Society, Cambridge, CB2 1ER, Engl. 1947-- . Three issues per year. ISSN 0022-1430.

A journal in the earth sciences oriented toward frozen environments (glaciers, sea ice, lake ice, permafrost, high altitude snow, and other ice situations). The focus is primarily on the physics, chemistry, and mechanics of ice and snow features and environments. Contains an excellent bibliography of the world's literature in these subjects. Vol. 15, no. 73 (1975), is composed of papers presented at the Symposium on Remote Sensing in Glaciology held at Cambridge, England, on 16-20 September 1974. (See citation no. 165.)

348. JOURNAL OF SPACECRAFT AND ROCKETS. American Institute of Aeronautics and Astronautics, 1290 Avenue of the Americas, New York, N.Y. 10019. 1964-- . Monthly. ISSN 0022-4650.

This journal is primarily oriented toward hardware, technology, and engineering having occasional articles of direct interest to remote sensing for earth resources applications. A statement describing the scope of the journal states: "...devoted to the advancement of the science and technology of space flight through . . . papers describing significant advances in space technology, the use of spacecraft and the applications of space technology in other fields."

349. JOURNAL OF WATER POLLUTION CONTROL. Formerly SEWAGE WORKS JOURNAL (to 1950), and SEWAGE INDUSTRIAL WASTES (to 1960). Water Pollution Control Federation, 2626 Pennsylvania

Ave., N.W., Washington, D.C. 20037. 1928-- . Monthly. ISSN 0043-1303. Microform copies available.

The Water Pollution Control Federation, the publisher of this journal, is composed of thirty-nine member associations in the United States and affiliated with twenty-one organizations in other countries. This journal occasionally contains papers dealing with remote sensing of water pollution features such as sewage, effluents, and turbidity. Summaries are in English, French, German, Portuguese, and Spanish. Cumulative indexes are available. It contains figures, tables, references, and advertisements. "Occasionally this journal is listed as WATER POLLUTION CONTROL FEDERATION - JOURNAL).

349A   LANDSAT DATA USERS NOTES. U.S. Geological Survey. EROS Data Center, Sioux Falls, S.Dakota 57198. 1978-- . Bimonthly.

This is possibly one of the most timely publications (free) for the user of Landsat data to receive. Because the majority of the Landsat data used is obtained from the EROS Data Center, this publication is most helpful in documenting policy changes, new products, training schedules standard catalogs, and other items of interest. The bimonthly schedule was initiated with no. 3 (November 1978) and at that time it was planned that the issues would concentrate on topics of interest to the remote-sensing data users with the emphasis on Landsat-related developments.

350.   MARINER'S WEATHER LOG. U.S. Naval Oceanographic Data Center, Page Building 1, Room 400 D-762, Washington, D.C. 20235. 1957-- . Bimonthly. ISSN 0025-3367. Available on microfiche.

This publication, free to mariners, is basically a climatic review of the North Atlantic, North Pacific, and the Great Lakes areas. It is an excellent source of information concerning the weather and meteorological conditions of these maritime areas and often contains imagery of cloud and severe weather situations from meteorological satellites. It contains charts, illustrations, figures, and references together with smooth and rough logs of the weather of previous months.

351.   MODERN GEOLOGY. Gordon and Breach Science Publishers, Ltd., 42 William IV Street, London WC 2, Engl. 1969-- . Four issues per year. ISSN 0026-7775. Available on microfiche.

Although this publication deals specifically with the field of geology, there have been several articles dealing with remote-sensing applications and interpretative techniques published. Several publications concerning radar imagery and IR properties (reflectance) of various minerals and rock types are especially noteworthy. The journal contains, in addition to scientific articles, advanced book reviews.

352.   MONTHLY WEATHER REVIEW. American Meteorological Society, 45 Beacon St., Boston, Mass. 02108. 1872-- . Monthly. No ISSN.

This journal is a medium for the publication of research on meteorological topics and publishes papers in which the emphasis of the research is on the solution of operational problems related to weather observations, instrumentation and weather analysis, and forecasting. This is an excellent source for research dealing with meteorological satellites such as Nimbus, SMS, TIROS, and so forth.

352A   NASA REPORTS TO EDUCATORS. Educational Services Branch, Public Affairs Division (LFG-13), National Aeronautics and Space Administration, Washington, D.C. 20546. 1971-- . Quarterly.

This (free) publication is prepared for the community of educators, especially at the elementary and secondary school level. This publication is quite important for this group because it provides listings of publications, films and services available to educators, and the addresses from which to request them. Also short news items giving brief backgrounds of new developments of interest to the identified group of educators are included.

353.   New Mexico, University of. Technology Application Center. QUARTERLY LITERATURE REVIEW OF THE REMOTE SENSING OF NATURAL RESOURCES. Albuquerque: 1974-- .

For annotation see item no. 305.

354.   OIL AND GAS JOURNAL. Petroleum Publishing Co., 1421 S. Sheridan Rd., Tulsa, Okla. 74101. 1902-- . Weekly. ISSN 0030-1388.

This industry-oriented journal deals primarily with information concerning political, economic, and logistical aspects of the oil and gas industry. Occasionally remote-sensing articles of special interest to the industry will be included. Also available on microfiche from Xerox University Microfilms.

355.   PATTERN RECOGNITION. Pergamon Press, Inc., Journals Department, Maxwell House, Fairview Park, Elmsford, N.Y. 10523. 1968-- . Quarterly. ISSN 0031-3203. Available on microfilm.

Because much of the interpretation which is conducted on remotely sensed data is of a quantitative nature, and because this aspect of interpretation is still very much in its infancy, techniques of pattern recognition and the developments in this field are especially important to the remote-sensing researcher. Thus this journal of the Pattern Recognition Society is an excellent source for articles dealing with these techniques and advances. The journal also contains book reviews.

356.　PHOTOGRAMMETRIA. Elsevier Scientific Publishing Co., Box 211, Amsterdam, Netherlands. 1949-- . Bimonthly. ISSN 0031-8663.

The official journal of the International Society for Photogrammetry, this European journal contains articles of special interest to photogrammetrists and photo interpreters. Individual issues often contain two or three articles, many of which are quite detailed and complete. Articles are published in English, French, and German. It contains illustrations, charts, figures, references, and some advertisements.

357.　PHOTOGRAMMETRIC ENGINEERING AND REMOTE SENSING. American Society of Photogrammetry, 105 North Virginia Avenue, Falls Church, Va. 22046. 1975-- . Monthly. ISSN 0031-8617.

This standard and excellent American journal dealing with photogrammetry, photo interpretation, and remote-sensing interpretations and applications is the successor to PHOTOGRAMMETRIC ENGINEERING (1934-75). Normally the journal contains eight to ten articles, numerous advertisements for remote-sensing products and services, announcements of upcoming publications and symposia, and several book reviews. Illustrations, article abstracts, and some letters from the readers are included.

358.　PHOTOGRAMMETRIC RECORD. Photogrammetric Society, 47 Tothill St., London, SW1H 9LH, Engl. 1953-- . Semiannual. ISSN 0031-868X.

As the name implies, this journal is oriented primarily toward photogrammetry rather than photo interpretation, and therefore contains numerous articles dealing with mensuration and its history. Book reviews, abstracts, bibliographies, figures, charts, and illustrations are included in the journal. As more instruments and techniques become available for making measurements from remote locations, it can be expected that the journal will continually expand to include within its pages papers concerning these systems and their output.

359.　PROCEEDINGS IN PRINT. P.O. Box 247, Mattapan, Mass. 02126. 1966-- . Six issues per year. ISSN 0032-9568.

This is an index to conference proceedings in all subject
areas and in all languages. Included are reports or proceedings of conferences, symposia, lecture series, con-
gresses, hearings, seminars, courses, institutes, colloquia,
meetings, and published symposia. Each conference appears
under its unique title, and, wherever possible, place, data,
and sponsorship of the conference, together with ordering
information are included. Remote sensing is one of the
keywords in the index. Back issues are available from the
publisher.

360. RADIO SCIENCE. American Geophysical Union, 1909 K St., N.W.,
Washington, D.C. 20006. 1966-- . Monthly (except September).
ISSN 0048-6604.

The International Union of Radio Science (Union Radio Scientific International, URSI) was founded in 1919 as a medium for worldwide cooperation in radio research. The aims
of the union are to promote the scientific study of radio
communications, to aid and organize radio research on an
international scale and to facilitate agreements upon common
methods of measurement as well as standardization of measuring instruments. The U.S. National Committee of URSI
is a committee of the National Academy of Science and the
National Research Council, which coordinates activities of
American scientists and engineers in the furtherance of these
objectives. This journal is the major place of publication
for the results of these efforts. Because so much of remote
sensing deals with the transmission of various types of data
via radio communication techniques, many of the articles
are of direct interest.

361. REMOTE SENSING OF ENVIRONMENT: AN INTERDISCIPLINARY
JOURNAL. American Elsevier Publishing Co., 52 Vanderbilt Ave.,
New York, N.Y. 10017. 1969-- . Quarterly. ISSN 0034-4257.

This journal ". . . serves the diverse remote-sensing community
with the publication of scientific and technical results on
theory, experiments and systems design in remote-sensing
technology and applications. In addition to original research papers, surveys and summaries of previously piece-
wise published works are welcome, as are comprehensive
state-of-the-art articles. Tutorial papers of interest to teachers
and those engaged in remote-sensing training are acceptable
if sufficiently broad in their coverage. Brief papers containing little or no analysis but presenting new data will be
published as Short Communications."

362. REMOTE SENSING OF THE ELECTRO-MAGNETIC SPECTRUM. Department of Geography-Geology, University of Nebraska at Omaha,

Omaha, Neb. 68101. 1973-- . Quarterly. No ISSN.

This small journal is the forum of the Remote Sensing Committee of the Association of American Geographers. As such it contains short articles concerning remote-sensing applications, sources of data, letters, book reviews, and general commentary about remote sensing. It is generally oriented toward the academic community. Several recent issues have been devoted to special topics, for example, radar remote sensing and educational methods in remote sensing. Beginning January 1979, this journal appeared as REMOTE SENSING QUARTERLY.

363. SCIENCE. American Association for the Advancement of Science, 1515 Massachusetts Ave., Washington, D.C. 20005. 1880-- . Weekly. ISSN 0036-8075.

Some of the major objectives of the AAAS are to foster the work of scientists and to improve the effectiveness of science in the promotion of human welfare. This journal, in keeping with such objectives, occasionally publishes articles pertaining to remote sensing of earth and environmental problems. Radar observations of hurricanes, infrared detection of atmospheric water vapor, and use of geostationary satellites have been the subjects of some reports in the 1977 volumes. In addition, longer pieces (termed "articles") during this year have included a review of Landsat processing via computers and a set of commentaries on Landsat work (see vol. 196, no. 4289, 29 April 1977).

364. SCIENTIFIC AND TECHNICAL AEROSPACE REPORTS (STAR). Scientific and Technical Information Facility, National Aeronautics and Space Administration, Washington, D.C. 20546. 1963-- . Biweekly. ISSN 0036-8741. Individual copies and the cumulative indexes are available from the Superintendent of Documents, U.S. Government Printing Office, Washington, D.C. 20402. Available in both hard copy and microfiche versions through NTIS.

This publication covers reports in the literature of the science and technology of space and aeronautics. All such NASA reports, together with similar material from other government agencies, universities, corporations, and research organizations are included. The papers are grouped under a number of headings and are cross-referenced by subject, author, contract number, NASA accession number, and corporate source. Annual cumulative indexes are available. Prior to 1963 this series had several titles, formerly NASA TECHNICAL PUBLICATION ANNOUNCEMENTS and INDEX OF NASA TECHNICAL PUBLICATIONS. Informatics Information Systems Company presently prepares this journal for NASA.

365. SPACEFLIGHT. British Interplanetary Society, Ltd., 12 Bessborough

Gardens, London, Engl. SW1V 2JJ. 1956-- . Monthly. ISSN
0038-6340.

> Although of limited usefulness to the earth-resources com-
> munity, this journal does present the latest developments in
> satellites. The "milestone" section, giving important dates
> and a synopsis of events is often helpful for chronological
> information. It covers all countries and developments, but
> is especially good for Soviet developments.

366. U.S. National Aeronautics and Space Administration. Scientific and
Technical Information Office. EARTH RESOURCES: A CONTINUING
BIBLIOGRAPHY WITH INDEXES. NASA SP-7041. Washington, D.C.:
U.S. Government Printing Office, 1974-- . Available from NTIS.

> For annotation see item no. 313.

367. WATER RESOURCES BULLETIN. American Water Resources Association,
Water Resources Bulletin Editorial Offices, 508 S. Sixth St., Cham-
paign, Ill. 61820. 1965-- . Bimonthly. ISSN 0043-1370.

> A journal of water resources research, planning, develop-
> ment, and management, this journal includes assorted arti-
> cles on the use of remote sensing in these study areas. Of
> special interest is vol. 10, no. 5 (October 1974), which
> contains five articles on satellite analysis of the 1973 Mis-
> sissippi River floods. The journal contains abstracts, book
> reviews, charts, illustrations, and references.

Chapter 8

WORKSHOPS, UNIVERSITY, AND TRAINING COURSES

This section consists of a listing of courses, workshops, and seminars which were available during the end of 1977 and early 1978, and those which were projected for the coming year. This listing is based upon the available brochures and prospectus prepared by the organizers as well as a series of conversations with instructors. These courses change rapidly but there are still several methods by which the interested individual can maintain an awareness of their growth, development, and demise.

One can correspond with the organizations listed. However, when new organizations and individuals appear to offer short courses and other instruction, this method is of little merit. To identify those who are entering the field, one should scan the journals dealing with remote sensing. Those of particular importance in this area are the following: PHOTOGRAMMETRIC ENGINEERING AND REMOTE SENSING, GEOTIMES, CANADIAN JOURNAL OF REMOTE SENSING, REMOTE SENSING OF THE ENVIRONMENT, LANDSAT DATA USERS NOTES, JOURNAL OF THE BRITISH INTERPLANETARY SOCIETY, ITC JOURNAL (cited in journal section).

368.   Association of American Geographers, 1710 Sixteenth St., Washington, D.C. 20009.

> For the past five years, the Association of American Geographers have conducted a one- to two-day remote-sensing short course in conjunction with their annual meeting. This meeting is generally held in March or April at various locations within the United States and Canada. The remote-sensing short courses tend to be tutorial and emphasize introductory training in remote sensing--consisting of lectures and some "hands-on" experience with remote-sensing data, primarily imagery. These courses are, however, organized on a annual basis and are, therefore, not a regularly scheduled part of the program. Each year, the association solicits proposals from the membership to conduct sessions for presenting papers as well as various training groups and workshops, and it is through this mechanism that the remote-

sensing short courses are developed. Generally, in October or November the association has a relatively firm commitment with respect to the upcoming annual meeting, and it is at this time that one should contact the association to learn of the projected meetings.

368A   Bidwell, Timothy C. "College and University Sources of Remote Sensing Information." PHOTOGRAMMETRIC ENGINEERING AND REMOTE SENSING 41 (October 1975): 1273-84.

This paper is essentially a listing of remote-sensing newsletters, institutes, laboratories, and programs which were identified, via a mail questionnaire, in the fall of 1974. Although, at present, about five years out of date, it is one of the most complete sources available for this type of information, and many of the groups identified are still in operation. A subject index (by academic discipline) and a geographic index (by state) is quite helpful.

369.   California, University of. Berkeley. Continuing Education in Environmental Design, University Extension, 2223 Fulton St., Berkeley, Calif. 94720. (415) 642-4811.

In April 1978, a three-day short course titled "Remote Sensing: Its Application to Natural Resource Management" was presented at Berkeley, Calif. According to the available brochure: Course material and workshops focus on the physical principles that underlie the technology, sensor system, formats, and data processing techniques used in remote sensing. Commonly available imagery types were analyzed and problems dealing with application to agriculture, forestry, soil science, and range, recreation, watershed, and wildlife management studied in workshop sessions.

370.   Earth Resources Observation Systems (EROS) Data Center, Applications Assistance Branch, EROS Data Center, Sioux Falls, S.Dak. 57198. (605) 594-6511, ext. 111.

The EROS data center provides a variety of training and assistance programs to individuals and groups interested in working with remotely sensed data. The Applications Assistance Branch provides state-of-the-art remote-sensing capabilities to personnel within the USDI and to other cooperating agencies, foreign countries, and general users. Orientation sessions, of two-to four-days duration provide the basic principles and techniques. Disciplinary or techniques workshops provide training on a more advanced level. A structured course titled the International Remote Sensing Workshop, lasting four weeks and given semiannually is oriented to foreign scientists and resource managers. This course is the only one given on a routine schedule, whereas other courses are designed, announced, and presented in response to a defined need by the users. For descriptions of specific programs presented at any given time the organization should be contacted directly. It should also be contacted to determine if a course can be offered to fit a particular users' need.

371.   George Washington University, Continuing Engineering Education, George Washington University, Washington, D.C. 20052.

The school of Engineering and Applied Science, under their Continuing Engineering Education program, sponsors a series of short courses. Two of these, most recently presented, are titled Remote Sensing and Digital Information Extraction and Digital Image Processing of Earth Observation Sensor Data. The former, course no. 488, "...has been designed for those involved in converting remotely sensed data into information products. In addition to presenting these techniques, a technology assessment will discuss current and future sensor, on-board processing, and general data processing trends." The companion course, no. ?07 "...is designed for physical scientists, photo interpreters, system analysts, engineers, and programmers who need a better knowledge of advanced sensors and digital image correction enhancement and information extraction techniques." Each of these courses consists of five days of instruction and are designed to complement one another.

372.   Indiana State University, Dr. Paul Mausel or Dr. Samuel Goward, Indiana State University Remote Sensing Laboratory, Department of Geography and Geology, Terre Haute, Ind. 47809.

"Through the support of a grant from the National Science Foundation, the staff of the Indiana State University Remote Sensing Laboratory (ISURSL), in the Department of Geography and Geology, Indiana State University, is presenting a series of short instructional programs in applications of remotely sensed data. The program is specifically designed to introduce in-service scientists, university faculty, and students involved in urban and regional planning, environmental analysis, and mineral resources development, to state-of-the-art remote sensing techniques that will benefit their studies. The courses [are] presented either on the ISU campus or at a participant-group selected location. A diverse number of course offerings are available, including 1, 2, 5, and 10 day instructional programs." (From PHOTO-GRAMMETRIC ENGINEERING AND REMOTE SENSING 44, no. 4, April 1978, p. 468.) These courses will be conducted between October 1978 and September 1979.

373. Integrated Computer Systems, Inc. Integrated Computer Systems, Inc., 3304 Pico Blvd., P.O. Box 5339, Santa Monica, Calif. 90405. (213) 450-2060.

> This company presents a series of very intensive short courses, generally lasting for about three days, at several locations within North America. Generally Washington, D.C., Los Angeles, and Toronto (in addition to other locations) have been selected. The courses are presented by engineers and are oriented to the engineering community interested primarily in hardware development and technical knowledge of equipment operation and design. These courses are possibly, therefore, of limited utility to the earth-resources community. Courses presented in the past have included Digital Signal Processing, Computer Graphics, and Synthetic Aperture Radar (SAR) Systems.

374. Laboratory for Applications of Remote Sensing. D.B. Morrison, Purdue University, West Lafayette, Ind. 47907. (317) 749-2052.

> The Laboratory for Applications of Remote Sensing (LARS) at Purdue University has been one of the leading centers in the development of analysis techniques (especially for visible and infrared data), and also for their analysis, primarily in the areas of vegetation, crop, and water studies. They have conducted a series of symposia and short courses in these areas and have a well-developed self-instruction program consisting of programmed texts, slides, and tape decks. Two of their offerings are as follows:

> 1. Short Course on Remote-Sensing Technology and Applications

>> This course is presented monthly by LARS and has a limited enrollment. The material presented is individually tailored to the group convened for the course and stresses both the "hands-on" experience and detailed instruction. Landsat data have, heretofore, been the major data source. Various analysis methods, basic concepts of remote sensing, introductory statements about pattern recognition, and a series of statistical and analytical seminars are included. The program also includes comments on the current research and methods for obtaining data. The program is described in some detail in an article by B.M. Lube and J.D. Russel, "A Short Course on Remote Sensing," PHOTOGRAMMETRIC ENGINEERING AND REMOTE SENSING 43, no. 3 (March 1978): 299-302.

> 2. Advanced Topics in the Analysis of Remote-Sensing Data

For prerequisites, contact: Prof. Philip H. Swain
LARS
1220 Potter Dr.
West Lafayette, Ind.
47906   (317) 749-2052

For registration, contact: LARS-Short Courses in
Advanced Topics in
Analysis of Remote
Sensing Data
Continuing Education
Business Office
Room 110, Steward Center
Purdue University
West Lafayette, Ind.
47907

This course treats advanced techniques in the numerical analysis of remote-sensing data and builds on
the basic pattern recognition methods implemented
in a variety of computer software programs. Therefore, the course is intended for individuals concerned
with numerical analysis of remote-sensing data and
who have knowledge and experience with the fundamentals of quantitative remote sensing. The brochure
available (for the course given in May 1978) notes
two addresses for further information.

375.   Nealey, L.D. "Remote Sensing/Photogrammetry Education in the
United States and Canada." PHOTOGRAMMETRIC ENGINEERING
AND REMOTE SENSING 43, no. 3 (March 1977): 259-84.

This article is the most recent and readily available listing
of courses, programs, projects, and textbooks in the field
of photogrammetry and remote sensing covering North America. Originally prepared for presentation at the Forty-sec-
ond Annual Meeting of the American Society of Photogrammetry (February, 1976) and updated for journal publication
to include Canadian institutions, this listing is accurate
through 1975. A total of 470 courses in the United States
and 64 in Canada are listed, together with a listing of texts
(and the frequency of their selection) and a list of engineering research projects in remote sensing and photogrammetry.
A very valuable paper, and a good initial point for those
wishing to seek university training in these fields. See update of this article in (1) PHOTOGRAMMETRIC ENGINEER-
ING AND REMOTE SENSING 44 (August 1978): 1043;
(2) 44 (November 1978): 1383.

376.   Rodriguez-Bejarano, D. "The Teaching of Photo-Interpretation and
Photogrammetry in the Field of Natural Resources." PHOTOGRAM-
METRIC ENGINEERING AND REMOTE SENSING 33, no. 3 (March
1977): 285-91.

The subject of remote sensing and photogrammetry in Mexico
and Central America is the thrust of this article. The state-
of-the-art and trends in these courses in the field of natural
resources in higher education is reported upon. Thirty-two
separate institutions are included together with a listing of
the equipment and staff available at each.

377. The South Dakota School of Mines and Technology, Dr. Charles Thielen,
Director of Continuing Education, South Dakota School of Mines and
Technology, Rapid City, S.Dak. 57701. (605) 394-2480.

In October 1978 the South Dakota School of Mines and
Technology, in cooperation with the EROS Data Center,
presented a Workshop in Applications of Geological Remote
Sensing to Mineral Exploration. The brochure available
states the workshop will be held near the Black Hills at
the South Dakota School of Mines and Technology in Rapid
City. The Workshop is designed for mineral industry personnel who are actively engaged in mineral exploration.
Instruction emphasized principles and concepts utilized in
the application of remote-sensing technology to mineral exploration problems. Workshop exercises involve the analysis
and interpretation of aircraft, spacecraft, and satellite imagery stressing the application of fundamental principles to
geologic problems. It is also noted that there are four days
of intensive instruction and an optional fifth day devoted to
case histories and a field trip.

378. Stanford University. Contact the Association of American Geographers
or the EROS Program of the USGS.

In June 1978, the First Annual Conference on Remote Sensing Educators (CORSE-78) [was] held at Stanford University
in collaboration with the USGS, EROS Program, the American Association of Geographers, and the American Society of
Photogrammetry. The presentation of formal papers and three
days of working sessions along with a trade show and data
products exhibit comprised the conference.

AUTHOR INDEX

This index includes authors, editors, and other contributors cited in the text. It is alphabetized letter by letter, and numbers refer to entry numbers.

A

Abbott, W.W. III 275
Abdel-Hady, Mohamed 1
Advisory Group for Aerospace Research and Development (AGARD) 148, 173
Agnew Tech-Trans 34
Alaska, Department of Economic Development 2
Alexander, Larry 3
Alföldi, T.T. 3A
Ambionics 4
American Association for the Advancement of Science 363
American Association of Petroleum Geologists 318
American Astronautical Society 157, 158, 178, 179, 184, 343
American Chemical Society 332
American Congress on Surveying and Mapping 319
American Geographical Society 53
American Geological Institute 227
American Geophysical Union 346, 360
American Institute of Aeronautics and Astronautics 140, 323, 341, 348
American Meteorological Society 352
American Society of Limnology and Oceanography 168
American Society of Photogrammetry 141, 141A, 142, 143, 193, 208, 228, 229, 231, 232, 357

American Water Resources Association 189, 367
Anderson, Frank W., Jr. 5
Anson, Abraham 143
Arctic Institute of North America 144, 322
Arizona, University of. Office of Arid Land Studies 145
Arkell, R. 6
Association of American Geographers 276, 320, 368
Avery, T. Eugene 7, 204

B

Bailey, C.H. 8
Baker, D.R. 283A
Baker, Simon 205
Barrett, Eric C. 9, 146
Barwis, John H. 240
Battrick, B.T. 147
Bay, Sally M. 10
Beatty, F.D. 11
Becker, Margaret A. 159
Belew, Leland F. 12
Berlin, G. Lennis 284
Berne, University of. Institute of Applied Physics 183
Bernstein, Ralph 13
Bidwell, Timothy C. 285, 368A
Bird, J. Brian 241

Author Index

Blackband, W.T. 148
Bock, Paul 149
Bodechtel, Johann 14
Books, E. James 315A
Booz-Allen Applied Research 23
Bowker, David E. 15
Bradford, W.R. 206
Branch, Melville C. 16
Bressanin, G. 17
Brevard County (Florida) School Board
    17A
British Interplanetary Society 327, 365
Brosius, Charles A. 17A
Bruno, Richard O. 310
Bryan, M. Leonard 286
Bursnall, W.J. 178

C

California, University of. Berkeley
    207, 369
California, University of. Los Angeles.
    Engineering and Science Extension
    151
California, University of. Santa
    Barbara. Dept. of Geology 96
California Institute of Technology.
    Pasadena. Jet Propulsion Laboratory
    150
Canada. Defence Research Board 152
Canada Centre for Remote Sensing 201,
    3A
Canadian Aeronautics and Space Insti-
    tute 3A, 153, 188, 329
Carter, William D. 136, 287
Center for Experimental Design and
    Data Analysis 6
CENTO Seminar 171
Chardon, R.E. 288
Chown, M.C. 241
Coate, Godfrey T. 83
Collins, W. Gordan 126, 154
Colvocoresses, Alden P. 18
Colwell, Robert N. 19, 208
Committee on Space Research (COSPAR)
    149
Cooper, Saul 155
Corian, Edward F. 209
Cortright, Edgar M. 20
Council of Planning Librarians 284,
    294, 297
Csallany, Sandor O. 189
Curtis, Leonard F. 9, 146

D

Dale, William J. 242
Darden, Lloyd 21
Davis, R. 22
Davis, Shirley M. 210
Deer, Albert J. 211
Dellwig, Louis F. 212, 289
Denny, Charles S. 242
Department of Trade and Industry
    (England) 206
Desaussure, Hamilton 174
Deutsche Gesellschaft fur Photogram-
    metrie 326
Dill, Henry W., Jr. 205
Dismachek, Dennis C. 243
Dow, Donald H. 242
Duc, N.T. 147

E

Eakins, Richard H. 2
Earth Satellite Corp. 23
East Anglia, University of. Norwich
    334
Eastman Kodak Co. 24, 204
Ecological Society of America 168
ECON, Inc. 25, 38
Ekimov, Roza 290
El-Baz, Farouk 50
El-Kassas, Ibrahim A. 1
Elliot Automation Space and Ad-
    vanced Military Systems, Inc. 26
Engineering Topographic Laboratories
    84, 91, 128, 193
Environmental Research Institute of
    Michigan 58, 176, 215, 245,
    286, 292
Estes, John E. 27
European Space Research Organization
    17, 147, 214
Ewing, Gifford C. 156
Eynard, Raymond A. 28

F

Ferdman, Saul 157
Fischer, William A. 29, 202
Flanders, Allen F. 283A
Fleming, M. 283A
Florance, Edwin T. 77

Ford, C. Quintin 158
Freden, Stanley 159, 160
Friedman, David B. 160

G

General Electric Co. Space Division
30, 213
Geological Society of America 325,
336
George Washington University. Continuing Engineering Education 371
Gervin, Jannette C. 17A
Giddings, L.E. 244, 244A, 244B
Gierloff-Emden, Hans-Gunter 14
Glasby, J.P. 291
Goldstein, Harris M. 310
Gonin, G.B. 31
Goodyear Aerospace 32, 90, 91
Grady, James 33
Greenblatt, E.J. 58
Greve, T. 160A
Grigor'yev, A.A. 34
Guyenne, T.D. 179A

H

Haack, Barry N. 292
Hanes, Thomas E. 184
Haralick, Robert M. 35
Harger, Robert O. 36
Harper, Dorothy 37
Hassel, Philip G. 245
Hazelrigg, George A., Jr. 38
Heaslip, George B. 39
Heiken, Grant 246
Heiman, Grover 39A
Heiss, Klaus P. 38
Hemphill, W.R. 29
Hewish, Anthony 40
Higham, A.D. 214
Holland, John W. 258
Holter, Marvin R. 41
Holz, Robert K. 42
Honea, Robert B. 293
Hood, V. 179A
Hoope, Eugene R. 247
Horton, Frank E. 43
Horvath, Robert 215
Howard, William A. 294, 297

Hudlow, M. 6
Hughes, J. Kenrick 15
Hundemann, Audrey S. 294A, 294B,
294C

I

Illinois, University of. Rock Mechanics Laboratory 102
Indiana State University. Remote
Sensing Laboratory 372
Institute of Electrical and Electronics
Engineers 161, 339, 340
Institute of Environmental Sciences
344
Instituto de Pesquisas Espaciais 162
Instrument Society of America 163
Integrated Computer Systems, Inc.
373
Interdepartmental Committee on Air
Survey 164
International Astronautical Federation
216
International Bank for Reconstruction
and Development 248
International Business Machine Corp.
13
International Geographical Union
190
International Glaciological Society
165, 347
International Institute for Aerial Survey and Earth Sciences (ITC)
218, 342,
International Remote Sensing Institute
217
International Society for Photogrammetry 166, 356
International Union for Forest Research
Organizations 44
Institute of Air and Space Law 174
Iowa Geological Survey 167
Israel Program for Scientific Translations
52, 139
Ivey, Nancy B. 295

J

Johnson, Philip L. 168
Joint Publication Research Service
169
Jones, Natalie E. 296

Author Index

K

Kahn, D. 214
Kansas, University of. Center Research in Engineering Sciences 212, 289, 295, 316
Karcht, James B. 297
Karegeannes, Carrie E. 131
Karr, Clarence, Jr. 45
Katz, Y.H. 170
Kennedy, J.M. 46, 47
Kleckner, Richard 238
Kock, Winston E. 48
Kong, J.A. 49
Kosofsky, L.J. 50
Kovaly, John J. 51
Kover, A. 29
Kroeck, Dick 249
Krumpe, Paul F. 298, 299
Kudritskii, D.M. 52

L

Lane, Robert K. 189
LaPrade, G.L. 91
Lawrence, Mary Margaret 171
Layton, J. Preston 172
Leberl, Franz 218
Lee, Willis Thomas 53
Leonardo, E.S. 91
Levine, Daniel 54
Lied, F. 160A
Lindenlaub, John C. 219, 220
Lintz, Joseph, Jr. 55
Llaverias, Rita K. 300, 301, 302
Lo, C.P. 56
Lockheed Electronics 244, 244A, 244B, 250
Lomax, John B. 173
Long, Maurice W. 57
Los Alamos Scientific Laboratory 246
Lowe, D.G. 291, 302
Lowe, D.S. 58
Lowman, Paul D., Jr. 59
Lube, Bruce M. 219
Lueder, Donald R. 60
Lull, Howard W. 82

M

McCullough, Thomas J. 200
MacDonald, Harold C. 212
McGill University 174

McGinnies, W.G. 303
Manji, Ashraf S. 304
Marshall, Ernest W. 61
Massachusetts Institute of Technology, Res. Lab of Electronics 49
Matte, Nicolas Mateesco 174
Matthews, Richard E. 175
May, John R. 250A
Mercanti, Enrico P. 138, 159
Michigan, University of 62, 176, 177, 221, 222, 223
See also University of Michigan
Miller, Victor C. 224
Miller, Calvin F. 224
Mitchell, Cheryl A. 285
Moore, Patrick 272
Morgenthaler, George W. 178, 179
Morra, Robert 179
Morrison, A. 241
Mutch, Thomas A. 63
Mutter, Douglas L. 251

N

National Academy of Sciences 64, 65, 66
National Conference of State Legislaturers 10
National Geographic Society 273
National Research Council 67, 68, 69
National Research Council of Canada 328
National Scientific Laboratories 70
National Technical Information Service (see also NTIS INDEX) 338
NATO (North Atlantic Treaty Organization) Advisory Group for Aerospace Research and Development 148, 173
Nealey, L.D. 375
Nebraska, University of. Omaha. Department of Geography-Geology 362
Nefedov, K.E. 71
Nevada, Bureau of Mines and Geology 274
Nevada, University of. Reno 182

Newhall, Beaumont 72
New Mexico, University of. Technology Applications Center 252, 253, 254, 305, 353
Nicks, Oran W. 73
Nunnally, Nelson R. 225, 306

O

Operations Research Society 179
Optical Society of America 321
Ordway, Frederick Ira III 74

P

Page, Robert N. 75
Pearson, Bradley R. 307
Petrillo, Anthony J. 233
Photogrammetric Society (England) 358
Pierce, John R. 76
Plessey Radar Research Centre (England) 214
Plevin, J. 179A
Popov, I.V. 52
Popova, T.A. 71
Porter, Ronald A. 77
Pouquet, Jean 78
Prentice, Virginia 293
Price, Alfred 79
Purdue University. Laboratory for Applications of Remote Sensing 181, 210, 219, 220, 373

R

Ragusa, James M. 17A
Rajarajeswari, I. 308
Rango, Albert 182
Rao, M.S.V. 275
Ray, Richard G. 80
Raytheon Autometric Corp. 11
Raytheon Co. 81, 226
Reeves, Robert G. 227, 228
Reifsnyder, William E. 82
Reintjes, J. Francis 83
Remote Sensing Society (England) 126, 154
Richter, Dennis M. 255
Rinker, Jack N. 84
Robinove, Charles J. 85
Rodriguez-Lejarano, D. 376

Romanov, E.A. 52
Romanova, Mariya A. 86
Rosenfeld, Azriel 87
Rudd, Robert D. 88
Ruechardt, Eduard 89
Ruiz, Abraham L. 247
Russell, James 220
Ryan, Philip T. 155
Rydstrom, Hubert C. 90, 91
Ryerson, R.A. 3A

S

Sabins, Floyd F., Jr. 91A
Saint Joseph, K.S. 92
Schanda, Erwin 183
Scherz, James P. 93
Schneider, William C. 184
Schwertz, E.I., Jr. 288
Senger, Leslie W. 27
Sérs, Sidney W. 308A
Shahrokhi, Farouz 185
Shepard, Francis P. 94
Short, Nicholas S. 95
Simonett, David S. 55, 96
Singh, R. 97
Skolnik, Merrill I. 98
Smith, H.T.U. 98A
Smith, John T., Jr. 229
Smith, William L. 99
Society of American Foresters 345
Society of Photographic Scientists and Engineers 28
Society of Photo-Optical Instrumentation Engineers 170
South Dakota School of Mines and Technology 377
Space-General Corp., Advanced Microwave Systems Division 46, 47
Spurr, Stephen H. 100
Stafford, Donald B. 309, 310
Stanford University. School of Engineering 101, 378
Steiner, Dieter 311
Stevens, Alan R. 93
Strandberg, Carl H. 230
Summers, R.A. 58
Systems Control 22
Szuwalski, Andre 256

Author Index

T

Tandberg, E. 160A
Tarkoy, Peter J. 102
Tarnocai, C. 186
Technicolor Graphics 285, 312A
Tennessee, University of 185, 298
Tensor Industries 299
Texas A & M University 187
Theon, John S. 275
Thompson, G.E. 188
Thompson, Morris M. 231
Thompson, William I. III 312
Thomson, Keith P.B. 189
Thrower, Norman J.W. 276
Todd, William J. 312A
Tomlinson, R.F. 190
Transportation Systems Center 312
TRW Systems Group 103, 104
Twomey, C. 104A

U

Underwood, Richard W. 257, 258
UNESCO (United Nations Educational, Scientific and Cultural Organization) 191
United Nations 192
U.S. Air Force Cambridge Res. Labs 98A
U.S. Army. Corps. of Engineers (see also: Engineering Topographic Laboratories) 11, 84, 105, 128, 193, 240, 256, 296, 297, 309, 315A
U.S. Congress. House of Representatives 106, 107
U.S. Congress. Senate 108
U.S. Department of Agriculture 82, 205, 283
U.S. Department of Commerce 52, 283A, 338
U.S. Geological Survey 18, 23, 43, 80, 85, 136, 202, 232, 233, 238, 242, 259, 260, 277, 278, 279, 280, 281, 282, 287, 291, 293, 300, 301, 302, 304, 306, 307, 317, 372
U.S. Geological Survey. EROS Data Center 317, 349A, 370
U.S. Geological Survey. Water Resources Division 300, 301, 302

U.S. National Aeronautics and Space Administration 5, 12, 15, 19, 20, 30, 34, 50, 62, 66, 71, 73, 77, 95, 103, 104, 105, 109, 110, 111, 112, 113, 114, 115, 116, 117, 118, 119, 120, 121, 122, 123, 124, 131, 155, 159, 160, 175, 182, 194, 195, 196, 197, 198, 200, 211, 213, 233, 234, 235, 236, 241, 250, 257, 258, 259, 261, 262, 263, 264, 265, 266, 267, 268, 269, 275, 300-304, 313, 314, 315, 331, 364
U.S. National Aeronautics and Space Administration. Ames Research Center 22, 30, 31, 34, 263
U.S. National Aeronautics and Space Administration. Educational Programs Division 108, 110, 111, 112, 113, 114, 352A
U.S. National Aeronautics and Space Administration. Goddard Space Flight Center 197, 234, 265, 266, 267, 268
U.S. National Aeronautics and Space Administration. Johnson Space Center 104, 115, 116, 117, 175, 198, 235, 236, 250, 257, 258, 261, 262, 269
U.S. National Aeronautics and Space Administration. Scientific and Technical Information Division 118, 119, 120, 121, 122, 123, 154, 194, 196, 275, 313, 314, 315,
U.S. National Environmental Satellite Service 209, 243, 247, 270, 271
U.S. National Science Foundation 52, 71
U.S. Naval Oceanographic Data Center 350
U.S. Navy. Bureau of Ships 125
U.S. Navy. Office of Naval Research. Earth Science Division 47, 239
U.S. Soil Conservation Service 283
University of Michigan; Institute of Science and Technology Great Lakes Reserve Division 61

V

Van Genderen, John L.   126, 154
Verstappen, Herman Theodoor   126A
Veziroglu, T. Nejat   199
Vinogradov, B.V.   127
Vogel, Theodore C.   128, 315A
Von Bandt, Horst F.   128A
Von Fristag Drabbe, C.A.J.   237
Von Puttkamer, Jesco   200

W

Waite, William P.   212
Walters, Robert L.   316
Wanless, Harold R.   94
Warren, Charles R.   242
Water Pollution Control Federation
   349
Water Resources Information Center
   309
Watson-Watt, Sir Robert   129
Way, Douglas S.   130
Welch, Robin I.   2
Wells, Helen T.   131
Wenderoth, Sandra   132
Westerlund, Frank V.   317
Westinghouse Electric Co.   133

Wheeler, Gershon J.   134
White, Dennis   201
White, Leslie P.   135
Whiteley, Susan H.   131
Widger, William K., Jr.   135A
Wiedel, Joseph W.   238
Wilkinson, B.   214
Williams, Richard S., Jr.   136
Wisconsin, University of.   Remote
   Sensing Program   93, 97
Wolf, Paul R.   137
Wolfe, William L.   239
Wolff, Edward A.   138
Woll, P.W.   202
Woods Hole, Massachusetts
   Oceanographic Institution   156
Wyoming, University of.   Depart-
   ment of Geology   330

Y

Yost, Edward   132

Z

Zdanovich, V.G.   139
Zirkind, Ralph   203

TITLE INDEX

This index includes titles to all books, essays, maps, and journals cited in the text. It is alphabetized letter by letter. Numbers refer to entry numbers.

A

Active Microwave Workshop Report 175

Advanced Scanners and Imaging Systems for Earth Observations 195

Advanced Techniques for Aerospace Surveillance 148

Aerial Discovery Manual 230

Aerial Photographic Interpretation: Principles and Applications 60

Aerial Photographs in Geologic Interpretation and Mapping 80

Aerial Photography: The Story of Aerial Mapping and Reconnaissance 39A

Aerial Photography and Photo Interpretation 237

Aerial Photography and Remote Sensing for Soil Survey 135

Aerial Photography Used in Mapping Vegetation and Soils 127

Aerial Remote Sensing: A Bibliography 309

Aerial Surveys and Integrated Studies, Proceedings of the Toulouse Conference 191

Aerogeology 128A

Airborne Camera: The World from the Air and Outer Space 72

Airborne Instrumentation Research Project, Summary Catalogs 261

Air Photo Interpretation of Great Lakes Ice Features 61

Air Survey of Sand Deposits by Spectral Luminance 86

Alaska Remote Sensing Symposium (1964) 2

All-Digital Precision Processing of ERTS Images 13

AMERICAN ASSOCIATION OF PETROLEUM GEOLOGISTS. BULLETIN (journal) 318

AMERICAN CARTOGRAPHER (journal) 319

ANNALS OF THE ASSOCIATION OF AMERICAN GEOGRAPHERS (journal) 320

Annotated Bibliography and Evaluation of Remote Sensing Publications Relating to Military Geography of Arid Lands, An 303

Annotated Bibliography of Aerial Remote Sensing in Coastal Engineering, An 310

Annotated Bibliography of Bibliographies on Photo Interpretation and Remote Sensing 311

Annotated Bibliography of Remote Sensing Applied to Urban Areas 1950-1971, An 288

Annotated Bibliography of Reports, Studies and Investigations Relating to Satellite Hydrology 283A

Title Index

Annotated Bibliography of USGS Technical Letters--NASA Papers on Remote Sensing Investigations through June 1967 287

Apollo 9 Synoptic Photography Catalog 252

Apollo 6 and 7 Synoptic Photography Catalog 253

Apollo 6 Photomaps of the West-East Corridor from the Pacific Ocean to Northern Louisiana 277

Application of Remote Sensing Techniques to Inter and Intra Urban Analysis, The 43

Applications of Remote Sensing, Aerial Photography and Instrumented Imagery Interpretation to Urban Area Studies 297

Applications of Remote Sensors in Forestry 44

Applications Review for a Space Program Imaging Radar 96

Applied Infrared Photography 24

APPLIED OPTICS (journal) 321

Approaches to Earth Survey Problems through Use of Space Techniques 149

ARCTIC (journal) 322

ASTRONAUTICS AND AERONAUTICS (journal) 323

Atlas of the Universe, The 272

Atlas of Urban and Regional Change --San Francisco Bay Region 282

Author Index to Published ERTS-1 Reports 285

Availability of Earth Resources Data 259

AVIATION WEEK AND SPACE TECHNOLOGY (journal) 324

B

BIBLIOGRAPHY AND INDEX OF GEOLOGY (journal) 325

Bibliography of Remote Sensing Applications for Planning and Administrative Studies 306

Bibliography of Remote Sensing of Earth Resources for Hydrological Applications, 1960-1967 300

Bibliography of Remote Sensing of Resources 296

Bibliography on Remote Sensing 308

Bildmessung und Lufbildwesen: Zeitschrift für Photogrammetrie und Fernerkundung 326

BRITISH INTERPLANETARY SOCIETY JOURNAL (journal) 327

C

CANADIAN JOURNAL OF EARTH SCIENCES (journal) 328

Canadian Journal of Remote Sensing 329

Canadian Symposium on Remote Sensing, Third 188

Capabilities of Remote Sensors to Determine Environmental Information for Combat 84

Catalog of Operational Satellite Products 247

Catalog of Tidal Inlet Aerial Photography 240

Catalogue of Satellite Photography of the Active Volcanoes of the World 246

CENTO Seminar on the Applications of Remote Sensors in the Determination of Natural Resources 171

City Planning and Aerial Information 16

Coastal Imagery Data Bank: Interim Report 256

Collation of Earth Resources Data Collected by ERIM Airborne Sensors 245

"Colleges and University Sources of Remote Sensing Information" 368A

Color: Theory and Imaging Systems 28

Conference on Scientific Experiments of Skylab 140

CONTRIBUTIONS TO GEOLOGY (journal) 330

D

Data Collection System: Earth Resources Technology Satellite-1 155

Data Preprocessing Systems for Earth Resources 17

Data Reduction of Airborne Sensor Records 81

Data Users Handbook--NASA Earth Resources Technology Satellite 213

Deciphering of Groundwater from Aerial Photographs 71

Definition of the Total Earth Resources System for the Shuttle Era (TERSSE) 30

Demeter: An Earth Resources Satellite System 101

Descriptive Catalog of Selected Aerial Photography of Geologic Features in the United States 242

Developing Earth Resources with Synthetic Aperture Radar 32

E

Earth from Space, The 14

Earth in the Looking Glass, The 21

Earth Observation from Space and Management of Planetary Resources 179A

Earth Photographs from Gemini VI through XII 118

Earth Photographs from Gemini III, IV, and V 119

EARTH RESOURCES: A CONTINUING BIBLIOGRAPHY WITH INDEXES (journal) 313, 331, 366

Earth Resources Mission Performance Studies 103

Earth Resources Program: Procedures Manual for Detection and Location of Surface Water Using ERTS-1 Multispectral Scanner Data 235

Earth Resources Program Synopsis of Activity 115

Earth Resources Research Data Facility Index 268

Earth Resources Research Data Facility R&D File 269

Earth Resources Satellites 108

Earth Resources Survey Benefit-Cost Study: Economic, Environmental, and Social Costs and Benefits of Future Earth Resources Survey Systems 23

Earth Resources Surveys from Spacecraft 105

Earth Resources Survey System 106

Earth Resources Technology Satellite 104

Earth Resources Technology Satellite A: Press Kit 109

Earth Science in the Age of the Satellite 78

Earth Survey Bibliography: A Kwic Index of Remote Sensing Information 312

Ecological Surveys from Space 120

Economic Evaluation of the Utility of ERTS Data for Developing Countries, An 58

Economic Value of Remote Sensing of Earth Resources from Space: An ERTS Overview and the Value of Continuity of Service, The 38

Electromagnetic Spectrum and Sound: 50 Modern Experiments, The 8

Electrons and Waves: An Introduction to the Science of Electronics and Communication 76

Elements of Photogrammetry: With Air Photo Interpretation and Remote Sensing 137

Environmental Data Handling 39

Environmental Remote Sensing: Applications and Achievements 146

Environmental Remote Sensing 2: Practices and Problems 9

Environmental Satellite Imagery: Key to Meteorological Records Documentation No. 5.4 270

ENVIRONMENTAL SCIENCE AND TECHNOLOGY (journal) 332

EREP Users Handbook 236

ERTS-A: A New Window on Our Planet 136

European Earth Resources Satellite Experiments 147

Evaluation of 1971 Remote Sensing Activity in Manitoba 186

Everyone's Space Handbook 249

Exploring Space with a Camera 20

Eye in the Sky: Introduction to Remote Sensing 37

F

Face of the Earth as Seen from the Air: A Study in the Application of Airplane Photography to Geography, The 53

Feasibility Study of Microwave Radiometric Remote Sensing 77

Final Report on State Use of Satellite Remote Sensing 10

Focus Series 1975: A Collection of Single-Concept Remote Sensing Educational Materials, The 210

Folio of Land Use in the Washington, D.C. Urban Area 278

Frequency Requirements for Active Earth Observation Sensors 70

FUNCTIONAL PHOTOGRAPHY (journal) 33

Fundamentals of Infrared Technology 41

Fundamentals of Remote Sensing (Collins and Van Genderen) 154

Fundamentals of Remote Sensing (University of Michigan) 221

G

GATE International Meteorological Radar Atlas 6

Gemini Photographs of the World: A Complete Index 244

Gemini Synoptic Photography Catalog 254

GEO ABSTRACTS-G: GEOGRAPHY, REMOTE SENSING AND CARTOGRAPHY (journal) 334

GEOFORUM: THE INTERNATIONAL MULTI-DISCIPLINARY JOURNAL OF PHYSICAL, HUMAN AND REGIONAL GEOSCIENCES (journal) 335

Geographical Applications of Aerial Photography 56

GEOLOGICAL SOCIETY OF AMERICA. BULLETIN (journal) 336

Geologic Exploration with High Resolution Radar 90

Geology of Mars, The 63

Geoscience Instrumentation 138

Geoscience Potentials of Side-Looking Radar 11

GEOTIMES: NEWS OF THE EARTH SCIENCES (journal) 337

"Glossary and Index to Remotely Sensed Image Pattern Recognition Concepts" 35

Government Reports Announcements and Index 338

Ground Systems for Receiving, Analyzing and Disseminating Earth Resources Satellite Data 216

Guidance for Application of Remote Sensing to Environmental Management: Appendix A: Sources of Available Remote Sensor Imagery 250A

H

Handbook of Military Infrared Technology 239

Handbook of Remote Sensing Techniques 206

Hartford, Conn., N.Y., N.J., Mass. (NK 18-9) (map) 279

Hydrographic Interpretation of Aerial Photographs 52

I

IEEE, TRANSACTIONS ON GEOSCIENCE ELECTRONICS (journal) 340

IEEE PROCEEDINGS (journal) 339

Impact of Space Science on Mankind, The 160A

Index Maps for Gemini Earth Photography 244A

Index to Landsat Coverage. Ed. 1 260

Infrared and Raman Spectroscopy of Lunar and Terrestrial Minerals 45

Instruments of Darkness 79

INTERNATIONAL AEROSPACE ABSTRACTS (IAA) (journal) 341

International Radar Conference 161

International Workshop on Earth Resources Survey Systems 194

Interpretation Manual for the Airborne Remote Sensor System 215

Interpretation of Aerial Photographs 7
Introduction to Basic Remote Sensing
for Engineering Geologists 102
Introduction to Electromagnetic Remote
Sensing with Emphasis on Applica-
tions to Geology and Hydrology,
An 227
Introduction to Quantitative Remote
Sensing, An 220
Introduction to Remote Sensing: The
Physics of Electromagnetic Radia-
tion 225
Introduction to Remote Sensing for
Environmental Monitoring, An 93
Introduction to the Mathematics of
Inversion in Remote Sensing and
Indirect Measurement 104A
ITC JOURNAL (journal) 342

J

JOURNAL OF ASTRONAUTICAL
SCIENCES (journal) 343
JOURNAL OF ENVIRONMENTAL
SCIENCES (journal) 344
JOURNAL OF FORESTRY (journal) 345
JOURNAL OF GEOPHYSICAL RE-
SEARCH (journal) 346
JOURNAL OF GLACIOLOGY
(journal) 347
JOURNAL OF SPACECRAFT AND
ROCKETS (journal) 348
JOURNAL OF WATER POLLUTION
CONTROL (journal) 349

K

Key to Meteorological Records Documentation Series No. 5.3 271

L

Landsat Data Users Notes 349A
Landsat Image Maps 280
Landsat Index Atlas of the Developing
Countries of the World 248
Landsat Non-U.S. Standard Catalog
263
Landsat U.S. Standard Catalog 264
Land Use in the Southwestern United

States--from Gemini and Apollo
Imagery (map) 276
Legal Implication of Remote Sensing
from Outer Space 174
Light: Visible and Invisible 89
Look of Our Land: An Air Photo
Atlas of the Rural United States,
The 205
Lunar Orbiter Photographic Atlas of
the Moon 15

M

Management and Utilization of Re-
mote Sensing Data: Proceedings
of the Symposium 143
Manual of Color Aerial Photography
229
Manual of Photogrammetry 231
Manual of Photographic Interpretation
208
Manual of Remote Sensing 228
Mapping with Remote Sensing Data
142A
MARINER'S WEATHER LOG (journal)
350
Matrix Evaluation of Remote Sensor
Capabilities for Military Geogra-
phic Information 128
Matrix of Educational and Training
Materials in Remote Sensing 219
Meeting of the Soviet-American
Working Group on Remote Sensing
of the Natural Environment from
Space 169
Meteorological Satellites 135A
Methods for Studying Ocean Currents
by Aerial Survey 139
Microwave Remote Sensing from
Space for Earth Resource Surveys
68
Military Thematic Mapping and Map
Compilation from Radar Imagery
91
Mission to Earth: Landsat Views the
World 95
MODERN GEOLOGY (journal) 351
Monitoring Earth Resources from Air-
craft and Spacecraft 19
MONTHLY WEATHER REVIEW (journal)
352

Moon as Viewed by Lunar Orbiter, The 50

Mosaics of ERTS-1 Imagery of Conterminous United States and Alaska (map) 283

Multispectral Photography for Earth Resources 132

Multispectral Scanning Systems and Their Potential Application to Earth Resource Survey 214

N

NASA Reports to Educators 352A

NASA Thesaurus 123

National Environmental Satellite Service Catalog of Products 243

Near Earth Photography from the Apollo Missions and the Apollo-Suyez Test Project 244B

Nimbus 5 Data Catalog 265

Nimbus 6 Data Catalog 266

Nimbus 6 User's Guide, The 234

O

Observing Earth from Skylab 110

Oceanography from Space: Proceedings 156

OIL AND GAS JOURNAL (journal) 354

Operational Applications of Satellite Snowcover Observations 182

Operational Products from ITOS Scanning Radiometer Data 209

Optical Instrumentation in Science, Technology and Society 170

Orders of Magnitude: A History of NACA and NASA 5

Origin of Radar, The 75

Origins of NASA Names 131

Our Changing Coastlines 94

P

Passive Microwave Measurements of Snow and Soil: A Study of the Theory and Measurements of the Microwave Emission Properties of Natural Materials 47

Passive Microwave Measurements of Snow and Soil 46

PATTERN RECOGNITION (journal) 355

Peaceful Uses of Earth Observation Spacecraft 62

Phoenix NI 12-7 Experimental 1:250,000 Scale Space Photomaps (map) 281

Photo-Atlas of the United States 33

Photogeology 224

PHOTOGRAMMETRIA (journal) 356

PHOTOGRAMMETRIC ENGINEERING AND REMOTE SENSING (journal) 357

PHOTOGRAMMETRIC RECORD (journal) 358

Photogrammetry and Photo-interpretation: With a Section on Applications to Forestry 100

Photography Equipment and Techniques: A Survey of NASA Developments 211

Photo Interpretation for Land Managers 204

Photo-interpretation Studies of Desert Basins in Northern Africa 98A

Pictoral Guide to Planet Earth 74

Picture Processing by Computer 87

Planning Challenges of the 70's in Space 179

Portrait USA: The First Color Photomosaic of the 48 Contiguous United States (map) 273

Practical Applications of Space Systems 65

"Practical Cataloging, Indexing, and Retrieval Systems for Remote Sensing Data, A" 97

Principles of Imaging Radars: An Intensive Short Course 222

Principles of Radar 83

Proceedings: Caltech/JPL Conference on Image Processing Technology Data Sources, and Software for Commercial and Scientific Applications 150

Proceedings; Conference on Remote Sensing in Arid Lands 145

Proceedings: NASA Earth Resources Survey Symposium 196

Proceedings, Seminar in Applied Remote Sensing 167

Proceedings: Symposium on Remote Sensing and Photo Interpretation 166

Proceedings in Print 180

Proceedings of a Seminar on Thickness Measurement of Floating Ice by Remote Sensing 152

Proceedings of the Commission on Geographical Data Sensing and Processing, Moscow, 1976 190

Proceedings of the Fall Technical Meeting 141

Proceedings of the First Annual William T. Pecora Memorial Symposium 202

Proceedings of the International Symposium on Remote Sensing of Environment 176

Proceedings of the Princeton University Conference on Aerospace Methods for Revealing and Evaluating Earth's Resources 172

Proceedings of the Second Seminar on Air Photo Interpretation in the Development of Canada 164

Proceedings of the Symposium on Electromagnetic Sensing of the Earth from Satellites 203

Proceedings of the Symposium on Remote Sensing in Marine Biology and Fishery Resources 187

Proceedings of Workshop for Environmental Applications of Multispectral Imagery 193

"Progress in Remote Sensing" 29

Propagation Limitations in Remote Sensing 173

Q

Quarterly Literature Review of the Remote Sensing of Natural Resources 305, 353

R

Radar, Sonar, and Holography: An Introduction 48

Radar Bibliography for Geoscientists 316

Radar Electronic Fundamentals 125

Radar Fundamentals 134

Radargrammetry 54

Radargrammetry for Image Interpreters 218

Radar Handbook 98

Radar Reflectivity of Land and Sea 57

Radar Remote Sensing for Geosciences: An Annotated and Tutorial Bibliography 286

Radar Remote Sensing for Geoscientists: Short Course Notes 212

Radiant Energy in Relation to Forests 82

RADIO SCIENCE (journal) 360

Remote Measurement of Pollution 121

Remote Sensing: A Better View 88

Remote Sensing: Energy-Related Studies 199

"Remote Sensing: Environmental and Geotechnical Applications" 3

Remote Sensing: Literature Search 290

Remote Sensing, 1969 217

Remote Sensing: Principles and Interpretation 91A

Remote Sensing: Techniques for Environmental Analysis 27

Remote Sensing and Highway Transportation: An Annotated Bibliography 292

Remote Sensing and the Earth 17A

Remote Sensing and Urban Analysis: A Project Supportative Bibliography 307

Remote Sensing and Water Resources Management 189

Remote Sensing Applications for Mineral Exploration 99

Remote Sensing Applied to Environmental Pollution Detection and Management: A Bibliography with Abstracts 294A

Remote Sensing Applied to Geology and Mineralogy: A Bibliography with Abstracts 234C

Remote Sensing Applied to Urban and Regional Planning: A Bibliography with Abstracts

Remote Sensing Bibliography for Earth Resources, 1966-67 301

Remote Sensing Bibliography for Earth Resources, 1968 302

Remote Sensing Bibliography for Earth Resources, 1969 291

Remote Sensing Data Processing 126

Remote Sensing for Resource and Environmental Surveys: A Progress Review--1974 64

Remote Sensing in Ecology 168

Remote Sensing in Heomorphology 126 126A

Remote Sensing in Hyphology: A Survey of Applications with Selected Bibliography and Abstracts 308A

Remote Sensing Laboratory Publication List, 1964-1976 295

Remote Sensing of Earth Resources (Shahrokhi) 185

Remote Sensing of Earth Resources (U.S. Congress) 107

Remote Sensing of Earth Resources: A Literature Survey with Indexes 313

Remote Sensing of Earth Resources: A Literature Survey with Indexes, 1970-1973 Supplement 315

Remote Sensing of Environment (California, University of) 151

Remote Sensing of Environment (Lintz and Simonett) 55

REMOTE SENSING OF ENVIRONMENT: AN INTERDISCIPLINARY JOURNAL (journal) 361

Remote Sensing of Environment--Workshop 207

Remote Sensing of Soil Moisture and Ground Water, Proceedings of the Workshop 153

Remote Sensing of Terrestrial Vegetation: A Comprehensive Bibliography 298

Remote Sensing of the Chesapeake Bay 197

REMOTE SENSING OF THE ELECTROMAGNETIC SPECTRUM (journal) 362

Remote Sensing of the Urban Environment: A Selected Bibliography 294

Remote Sensing Platforms 18

Remote Sensing Quarterly (journal) 362

Remote Sensing Research Project; Glossary of Aerial Photography and Remote Sensing in Geology and Earth Sciences 1

Remote Sensing with Special Reference to Agriculture and Forestry 67

Resource Satellites and Remote Airborne Sensing for Canada (Proceedings of the First Canadian Symposium on Remote Sensing) 201

Resource Sensing from Space: Prospects for Developing Countries 69

S

Satellite-Derived Global Oceanic Rainfall Atlas (1973 and 1974) 275

"Satellite Imagery Interpretations: Suggestions for Laboratory Design" 3A

Satellite Photomap of Nevada--1976 (map) 274

SCIENCE (journal) 363

SCIENTIFIC AND TECHNICAL AEROSPACE REPORTS (STAR) (journal) 364

Seasat Economic Assessment 25

Second Fifteen Years in Space: Proceedings of the Eleventh Goddard Memorial Symposium, The 157

Second Joint Conference on Sensing of Environmental Pollutants 163

Seeing Beyond the Visible 40

Selected Bibliography of Corps of Engineers Remote Sensing Reports 315A

Selected Bibliography of Remote Sensing 293

Selective Bibliography: Remote Sensing Applications in Land Use and Land Cover Bases, A 311A

Seminar on Space Applications of Direct Interest to Developing Countries 162

Side-Looking Radar Systems and

Their Potential Applications to Earth Resources Surveys 26

Side-Look Radar 133

Significant Accomplishments in Science 198

Skylab: A Preview of America's First Earth-Orbiting Space Station 111

Skylab, Our First Space Station 12

Skylab Earth Resources Data Catalog 267

Skylab Explores the Earth 116

Skylab Experiments 124

Skylab 4: Photographic Index and Scene Identification 257

Skylab Results, The 184

Skylab-3 Handheld Photography Alphabetized Geographical Feature List 250

Skylab 3: Photographic Index and Scene Identification 258

Space, Air and Photo Images for the Rocky Mountain States 251

Space Exploration and Applications 192

Spaceflight 365

Space for Mankind's Benefit 200

Space Panorama 59

Space Photography and Geologic Studies 31

Space Photography 1977 Index 262

Space Remote Sensing of the Earth Landscapes 34

Space Shuttle (U.S. NASA, Educational Programs Division) 112

Space Shuttle (U.S. NASA, Johnson Space Center) 117

Space Shuttle Payloads: Proceedings of the Symposium 178

Space Technology and Earth Problems 158

Specialists Meeting on Microwave Scattering and Emission from the Earth 183

Spectrum: There's More than Meets the Eye, The 113

Surveillant Science: Remote Sensing of the Environment, The 42

Survey and Analysis of Potential Users of Remote Sensing Data, Final Report: State Government Activities in Remote Sensing 4

Survey of Space Applications, A 122

Syllabus: Workshop on Remote Sensing and ERTS Image Interpretation 232

Symposium on Machine Processing of Remotely Sensed Data 181

Symposium on Remote Sensing in Glaciology, A 165

Symposium on Remote Sensing of the Polar Regions 144

Symposium on Significant Results Obtained from the Earth Resources Technology Satellite - 1 159

Synthetic Aperture Radar 51

Synthetic Aperture Radar Systems: Theory and Design 36

T

Technical Papers from the Annual Meeting 142

Terrain Analysis: A Guide to Site Selection Using Aerial Photographic Interpretation 130

Theory of Passive Remote Sensing with Microwaves: Final Report 49

Third Earth Resources Technology Satellite Symposium 160

This Island Earth 73

Three Steps to Victory: A Personal Account by Radar's Greatest Pioneer 129

Training Course on Data Reduction of Radar Topographic Imagery 226

U

University of Michigan Notes for a Program of Study in Remote Sensing of Earth Resources, The 177

Urban and Regional Planning Utilization of Remote Sensing: A Bibliography and Review of Pertinent Literature 317

Urban Applications of Remote Sensing 284

"Urban Photo Index for Eastern U.S." 255

Title Index

Useful Applications of Earth-Oriented Satellites 66

Use of Radar Images in Terrain Analysis: An Annotated Bibliography 289

Use of Remote Sensing in Conservation, Development, and Management of the Natural Resources of the State of Alaska, The 2

User Data Dissemination Concepts for Earth Resources, Final Report 22

Uses of Air Photography 92

Uses of Conventional Aerial Photography in Urban Areas: Review and Bibliography 304

Using Remote Sensor Data for Land Use Mapping and Inventory: A Users Guide 238

W

WATER RESOURCES BULLETIN (journal) 367

What's the Use of Land: A Secondary School Special Studies Project 233

Why Survey from Space? 114

World Atlas of Photography from TIROS Satellites I to IV 241

World Remote Sensing Bibliography Index, The 299

Worldwide Disaster Warning and Assessment with Earth Resources Technology Satellites 85

SUBJECT INDEX

This index includes main subject areas within the text. It is alphabetized letter by letter. Numbers refer to entry numbers.

A

Aerial photography   19, 92-93, 114,
    148, 177, 191
  bibliographies on   294C, 297,
    304, 307, 309, 310, 314
  in Canada   186, 188, 211
  cataloging and retrieval system
    for use with   97
  catalogs of tidal inlet photography
    240
  coastal studies and   94, 256, 310
  color film for   28
  course notes on   223
  in desert basin studies   98A
  forestry applications of   100, 127,
    166
  geographical applications of   53,
    56, 237
  geological applications of   80,
    94, 125, 224, 230, 242
  geomorphological applications of
    126A, 224
  glossary of terms used in   1
  groundwater studies and   71
  history of   72
  hydrological applications of   52,
    230, 309
  in industrial site location   255
  interpretation of   7, 60, 84, 164,
    166, 376

  bibliography of   311
  geological data and   237
  Great Lakes ice features and
    61
  hydrological data and   52
  manuals for   204, 208, 230
introduction to   137
journals about   326, 333, 345,
    356-58
manuals for   229
for mapping vegetation and soils
    127, 135, 146
of ocean currents   139
processing of data from   126
of the Rocky Mountain area   251
in urban and regional planning
    16, 43, 130, 255, 294C,
    297, 304, 307
by the U.S. Air Force   39A
See also Maps and atlases; Space
    photography
Aeronautics and aviation, journals
    about   323-24, 327, 341,
    343, 348, 364-65
Africa, North, photo-interpretation of
    desert basins in   98A
Agriculture
  applications of remote sensing
    and space technology to
    17A, 27, 55, 65-67, 105,
    120, 122, 179, 188, 194,
    196

atlases showing lands devoted to 276

bibliographies on 290, 298-99, 314, 316

ERTS program and 38, 136, 160

expected benefits of Landsat to 23

radar applications and 32

See also Soils; Vegetation

Airborne Instrumentation Research Project, catalog of data collected by 261

Airborne Remote Sensor System, interpretation manual for 215

Aircraft, thermal surveys by 29. See also Aerial photography

Air pollution. See Pollution

Alaska, applicability of remote sensing to 2

Alberta, Canada, remote sensing of marine resources and vertebrates in 149

Apollo-Soyuz Test Project, catalog of earth photographs from . 244B

Apollo space program

maps and atlases derived from 272, 276-77

photographic equipment and techniques of 211

photographs from 105, 272

catalogs of 244B, 246, 252-53

of earth 72, 74

of the Rocky Mountains 251

of volcanoes 246

Aquatic vegetation. See Marine biology; Vegetation, aquatic

Arctic region

application of remote sensing to 144, 188

application of SEASAT to 25

journals about 322

See also Alaska

Argentina, application of ERTS hydrologic data to 162

Arid lands. See Deserts

Arizona Ecological Test Site 145

Association of American Geographers, remote sensing training programs of 368, 378

Atlases. See Maps and atlases

Atmosphere

application of remote sensing and space technology to 6, 149, 158, 199

in the polar regions 144

ERTS program and 38

microwave sensing of 175

See also Meteorological phenomena; Weather and climate

Aviation. See Aeronautics and aviation

B

Biology

remote sensing of in the polar regions 144

space technology applied to 168, 192

See also Marine biology

Bolivia, application of ERTS data to 162

Brazil, application of ERTS data to 162, 216

Broadcasting, satellite applications for 66

C

California, aerial soil mapping of 166

California, University of, at Berkeley, remote sensing training course of 369

Canada

aerial photography in 164

ERTS program and 216

use of remote sensing in 37, 188, 201

water resource monitoring in 149, 189

See also Alberta, Canada; Manitoba, Canada; Saskatechewan, Canada

Cartography

aerial photography and 92

application of remote sensing and space technology to 9, 66, 105, 120, 122, 179A

bibliographies on 314, 316

ERTS program and 58, 136

journals about 319, 334
radar use in 32, 54, 91, 96
See also Maps and atlases
Central America. See Latin America
Chesapeake Bay, remote sensing of
197
City planning. See Urban planning
Clouds
aerial mapping of 6
bibliography on 283A
data available on 270
Skylab observations of 116
space photographs of 119
Coastal waters
application of remote sensing to
196
application of SEASAT to 25
bibliography on 310
index to aerial photographs of 256
See also Swamps, marine; Tidal
inlets
Coastlines, aerial photography of 94
Color photography
film, cameras, and components of
132
manual for 229
theory and history of 28
Communications
radar interpretation and mapping of
lines of 91
science of 76
space technology applied to 65–66,
158, 160A, 192
See also Broadcasting; Radio communications
Computers 170
bibliography on 293, 314
in earth resource monitoring and
processing 17, 29, 65, 126,
150, 177, 179A, 181
course notes on 223
in location of surface water 235
in side-looking radar systems 26
use of for the ERTS program 213
Corps of Engineers. See U.S. Army.
Corps of Engineers
Crops. See Agriculture; Vegetation

D

Dams, ERTS programs in the inspection
of 189

Data processing. See Computers
Demeter satellite system 101
Deserts
aerial photography and mapping
of 98, 127
bibliography on 303
remote sensing applied to 145
Skylab observation of 116
See also Sand
Developing countries
benefits of the space program to
192
economic value of ERTS data to
58
Landsat coverage of 248
prospects of remote sensing data
utilization by 69, 162
Disasters
assessment of using radar 32
ERTS monitoring of 85
Drainage
aerial photography of patterns of
130
mapping of Canadian patterns of
189
radar interpretation and mapping
of patterns of 91

E

Earth Observation Satellite 103
Earth resources. See Remote sensing
of earth resources; names
of specific resources (e.g.
Vegetation)
Earth Resources Aircraft Project.
See Airborne Instrumentation
Research Project
Earth Resources Experiment Package
(Skylab) 184
application of to Argentinian
studies 162
application of to European studies
147
data catalog from 267
users handbook for 236
See also Skylab
Earth Resources Mission Performance
Study 103
Earth Resources Observation System.
Data Center 108

training and assistance programs of 370, 377-78

Earth Resources Survey Program
bibliography from 287
data index from 268-69

Earth Resources Technology Satellite 64, 104
accomplishments, review, and applications of the project 5, 106, 108-9, 136, 159-60, 201
applied to European earth resource study 147
bibliography of reports based on 285
catalog of photographs of volcanoes from 246
color images from 28
data collection and processing from 155, 216
handbooks for 213
training in 232
disaster monitoring of 85
earth survey techniques of 149
economic value of 38
to developing countries 58
in ice measurement 165
maps and atlases derived from photographs of 283
photographs of the Rocky Mountain area from 251
processing of images from 13
in water resource management and location 189, 235
See also Landsat (satellite)

Ecology and environmental monitoring
aerial photography in 166
application of remote sensing and space technology to 17A, 65, 168, 194, 196-97, 199
bibliography on 314
ERTS program and 136, 160, 232
journals about 332, 344, 345, 349, 367

Electrical Industries Association.
Automatic Image Pattern Recognition Committee 35

Electromagnetic radiation, physics of 8, 89, 225

Electromechanical scanners 195

Electron-beam images 195

Electronics, science of 76

Environment. See Ecology and environmental monitoring

Environmental Research Institute of Michigan, collation of earth resources data of 245

Europe
cooperative earth resources program in 149
current and future earth observation programs of 179A
ERTS program data applied to 147

Europe, Western, impact of remote sensing on the development of 174

F

Farm land. See Agriculture

Federal government, natural resource related state programs of 10

Fisheries, application of remote sensing to 187

Fishing, ocean, application of SEASAT to 25

Floods
ERTS mapping of 189
forecasting of by remote sensing devices 153
satellite analysis of 85, 367

Forestry
aerial photography and mapping of 100, 127, 166
application of remote sensing and space technology to 44, 65-67, 82, 105, 120, 122, 147, 179, 188, 194
bibliographies on 290, 298, 314
ERTS program and 38, 136, 160
expected benefits of Landsat to 23
journals about 345
radar use in 32
radiant energy in relation to the study of 82
See also Tropical rain forests

G

Gas, natural. See National gas
  industry
Gemini space program
  maps and atlases derived from
      244A, 272, 276
  photographs from 20, 105, 118-
      19, 272
    catalogs of 254
    of earth 59, 72, 74
    of the Rocky Mountain area 251
    of volcanoes 246
Geodesy
  bibliography on 314
  radar applications in 96
  satellite applications in 66
Geography
  aerial photography in 53, 56
  application of remote sensing and
      space technology to 55,
      66, 105, 120, 122, 128,
      149-50, 190, 194
  bibliographies on 290, 293, 316
  information systems in 181
  journals about 320
  multispectral imagery in the study
      of 163
  space photography in 34, 119
  See also Topography
Geology
  aerial photography in 80, 94,
      128A, 130, 166, 224, 230, 237
    catalogs of 242
  application of remote sensing and
      space technology to 17A,
      27, 55, 66, 91A, 102, 105,
      120, 122, 147, 150, 179A,
      194, 196
  bibliographies on 290, 294B, 303,
      305, 314, 316
  ERTS program and 136, 160, 162
  interpretation of aerial data for
      169
  journals about 318, 325, 330,
      335-36, 351
  of Mars 63
  microwave measurements in 68,
      183
  notes for electromagnetic remote
      sensing in 227

radar applications in 32, 90
space photography in 31, 119
See also Rock formations; Vol-
    canoes
Geomorphology
  aerial photography in 94, 224,
      237
  application of remote sensing
      and space technology to 1
      126A, 147
    in mapping operations 27
    in the polar regions 144
Geophysics 153
  ERTS program and 136
  journals about 346
Geoscience
  bibliographies on 286, 289
  instrumentation systems used in
      138
  journals about 335, 340
Glaciology
  ERTS program and 149
  journals about 347
  remote sensing of in the polar
      regions 144
  satellite applications for 147
  symposium on remote sensing in
      165
  See also Ice
Global Atmospheric Research Program.
    (GARP) Atlantic Tropical
    Experiment 6
Goddard Memorial Symposium (11th)
    157
Government. See Federal government;
    State government
Great Lakes, interpretation of ice
    features of 61
Groundwater
  aerial photography and 71
  radar applications and 90
  remote sensing of 153

H

Harbors. See Ports and harbors
Hartford, Conn., satellite photo-
    graphs of 279
Hawaii, remote sensing of water re-
    sources in 189
High resolution pointable imager 103

Subject Index

Highways. See Transportation planning

Holland. See Netherlands

Holography 222
  fundamental principles of 48

Housing studies, air and space photography in 43

Hydrography
  interpretation of aerial photographs in 52
  radar interpretation and mapping in 91

Hydrology
  aerial photography in 230
  application of remote sensing and space technology to 55, 66, 105, 120, 122, 128, 147, 179A, 194
  bibliographies on 283A, 300-301, 308A, 314
  ERTS program and 149, 162
  Landsat program and 153
  notes for electromagnetic remote sensing in 227
  radar applications in 32
  See also Floods; Groundwater; Water resource planning; Watershed modeling

I

Ice
  bibliography on 283A
  evaluation of different sensors in the study of 81
  interpretation of Great Lakes 61
  microwave measurement of 30, 70, 183
  radar in the surveillance of 32
  symposium on remote sensing of 165
  thickness measurement of 152
  See also Glaciology

Improved TIROS Operation System, interpreting data from 209. See also Tiros satellites

Indiana State University. Department of Geography and Geology. Remote Sensing Laboratory, training courses of 372

Industrial site location, aerial photography index for 16, 255

Industry, economic value of the ERTS program to 38

Infrared photography
  application of 24
  interpretation of 84

Infrared radiometry, in aerospace surveillance 148

Infrared scanner, thermal, in ice studies 81, 165

Infrared spectroscopy, in mineral description and summary 45

Infrared technology
  application of 179A
  fundamentals and physics of 8, 41
  handbooks for 239

Integrated Computer Systems, Inc., remote sensing training courses of 373

Interferometer, applied to city planning studies 16

International Geographical Union, meeting (1972) 27

Iowa, remote sensing data problems in 167

Irrigation, application of remote sensing to 179A

K

Kansas, University of. Center for Research. Remote Sensing Laboratory, bibliography of publications by 295

Kawartha Lakes, vegetation mapping in 189

Kenya, ERTS application to rangeland studies in 58

L

Lake beds, playa, microwave measurement of 46

Landsat (satellite) 21, 29, 166, 174, 188, 202
  data user notes to 349A
  economic, environmental, and social costs of 23

geologic and mineralogic applications of 294B
hydrologic applications of 153
interpretation of photographs from 84
maps and atlases derived from 248, 273, 278, 280
objective generalizations of images from 9
photographs of earth from 74, 95
 indexes and catalogs to 260, 263-64
 of the Rocky Mountain area 251
 of the U.S. 33
 of volcanoes 246
snow observations of 18
See also Earth Resources Technology Satellite
Landsat Follow-On Program, proposed capabilities of 10
Land use planning
air and space photography in 43, 166, 294C, 304
application of remote sensing and space technology to 17A, 65, 146, 149, 196, 199, 238
bibliographies on 290, 293, 294C, 304, 312A
ERTS program and 38, 136, 232
expected benefits of Landsat to 23
maps and atlases for 205, 278, 282
multidisciplinary approaches to 233
multispectral imagery in 193
radar applications to 32
See also Regional planning; Urban planning
Lasers, use of in ice studies 81, 165
Latin America, impact of remote sensing on the development of 174. See also Argentina; Bolivia; Brazil
Light waves 89
 physics of 8, 132, 225
Lunar orbiter, photographs from 15, 20, 50

M

Magnetic recording 177

course notes on 223
Magnetometer, in city planning studies 16
Manitoba, Canada, remote sensing activities in 186
Maps and atlases
based on Apollo photographs 272, 276-77
based on Gemini photographs 244, 272, 276
based on Landsat photographs 273, 280
based on Skylab photographs 279
based on Tiros satellite photographs 241
of global oceanic rainfall 275
of land use 276, 282
lunar 15
of Nevada 274
of the San Francisco Bay Region 282
of the U.S. 33, 205, 273, 279, 283
of the universe 272
of Washington, D.C. 278
of the West-East corridor 277
See also Cartography
Marine biology, application of remote sensing to 187. See also Oceanography; Vegetation, aquatic
Marine resource planning
application of remote sensing and space technology to 17A, 65, 149, 196
ERTS program and 160
expected benefits to Landsat to 23
See also Oceanography, Fisheries
Marine resources, bibliography on 314
Mars, geology of 63
Massachusetts, satellite photograph map of 279
Medicine, space technology applied to 192
Meteorological phenomena
applications of remote sensing and space technology to 6, 149-50, 192

journals about 350
space photographs of 20, 119
space photography and 34
See also Atmosphere; Satellites,
meteorological; Weather and
climate
Microwaves 49, 177, 179A
course notes on 223
feasibility of for remote sensing
77, 93, 183, 217
physics of 8, 225
rationale for selection of frequencies
of 70
recent applications of 29, 68
use of to measure snow and soil
46
in water resource sensing 96
workshop report on 175
Millimeter radiometry, in aerospace
surveillance 148
Mineral resources
application of ERTS data to 58,
160
application of remote sensing and
space technology to 99, 122
bibliographies on 284B, 314
radar applications and 90, 96
spectroscopic analysis of 45
Mining, ocean, application of SEASAT
to 25
Mississippi River
ERTS mapping of the floods of 189
satellite analysis of the floods of
367
Moon
photographic selections of 15, 20,
50, 73
spectroscopic analysis of the
minerals of 45
Multispectral imagery, applications of
193
Multispectral scanner 13, 151, 166
in ice studies 165
potential applications of 214, 217
in soil mapping 135

N

National Aeronautics and Space Administration. See U.S.

National Aeronautics and
Space Administration
National Conference of State Legisla-
tures. Task Force on Uses
of Satellite Remote Sensing
for State Policy Formula-
tion, final report of 10
National Environmental Satellite
Service. See U.S. National
Environmental Satellite Ser-
vice
National Oceanographic and
Atmospheric Administration.
See U.S. National Oceano-
graphic and Atmospheric
Administration
Natural gas industry
application of SEASAT to 25
journals about 354
Natural resources. See Remote
sensing of earth resources;
names of specific resources
(e.g. Mineral resources;
Vegetation)
Navigation, satellite and space
technology applied to 66,
192
Netherlands, interpretation of geological photographs of 237
Nevada, satellite photograph map
of 274
New Jersey, satellite photograph map
of 279
New York, satellite photograph map
of 279
Nimbus (satellite)
background for understanding the
data from 77
catalogs to data acquired by
265-66
photographs of the earth from 74
user guide to data of 234

O

Oceanography
aerial mapping in 127
application of remote sensing and
and space technology to
55, 66, 105, 120, 122,

147, 149-50, 156, 158, 179, 194
  bibliography on 314
  ERTS program and 38, 136
  measurement of waves and currents in 70, 139
  microwave sensing in 70, 175
  of the polar regions 144
  See also Coastal waters; Mining, ocean; Swamps, tidal; Tidal inlets; Vegetation, aquatic; headings beginning with the the term Marine
Ocean transportation. See Shipping
Oil spills
  Airborne Remote Sensor System and 215
  bibliography on 294A
  microwave detection of 29
Optical-mechanical scanners 177
  course notes on 223
Optics, journals about 321

P

Panoramic camera, in ice studies 81
Pecora (William T.) Memorial Symposium, proceedings of 202
Permafrost, remote sensing of 144
Petroleum industry
  application of SEASAT to 25
  journals about 354
  radar applications in 90, 96
  See also Oil spills
Photography. See Aerial photography; Color photography; Infrared photography; Panoramic camera; Space photography
Physics
  of electromagnetic radiation 225
  of light 132
  of remote sensing 8
Planets, photographic selections of 73
Polar regions, symposium on remote sensing in 144. See also Alaska; Antarctic region; Arctic region
Pollution
  bibliography on 294A
  journals about 349

multispectral imagery in the study of 193
  remote sensing of 121, 163, 197, 217
  See also Oil spills
Population studies, air and space photography in 43
Ports and harbors, application of SEASAT to 25
Purdue University. Laboratory for Applications of Remote Sensing, symposia and training courses of 374

R

Radar 51, 151, 177
  in aerospace surveillance 148
  bibliographies on 286, 289, 316
  in city planning 16
  in the GARP program 6
  geological applications of 90
  geomorphological applications of 126a
  geoscience potentials of 11
  history and background of 79, 129
  in ice studies 81, 165
  interpretation of 84
    training courses on 212, 218, 226
  mapping applications of 91
    of soils 135
  reflectivity of signals of 57
  in remote sensing 26, 32, 54, 93, 96
  theory and design of and introductions to 36, 48, 75, 83, 98, 125, 129, 133-34, 161
    course notes and lectures on 212, 218, 222-23
Radiant energy
  in forestry studies 82
  use of to measure sand deposits 86
  See also Electromagnetic radiation
Radio communications, journals about 360
Radiometry, for remote sensing 177
  course notes on 223
  See also Infrared radiometry; Millimeter radiometry

Radio waves, physics of 8
Rainfall
    atlas of global oceanic 275
    bibliography on 283A
    microwave measurement of 70
Raman spectroscopy, in mineral de-
        scriptions and summaries 45
Rangeland, atlases showing 276
Rangeland management
    ERTS application to 136, 160
    in Kenya 58
    expected benefits of Landsat to 23
    space technology applied to 179
Regional planning
    air and space photography in 43,
        130
    bibliographies on 294C, 312A, 317
    development of geographical data
        for 190
    See also Rural planning; Urban
        planning
Remote sensing of earth resources
    bibliographies on 283A-317
    catalogs about 240-71
    general literature about 1-139
    glossary of terms used in 35, 123
    journals about 318-67
    legal aspects of 174, 179A
    manuals and guides to 204-39
    maps and atlases derived from
        272-83
    political implications of 179A
    proceedings about 140-203
    workshops, university and training
        courses on 368-78
Rice, ERTS applications to forecasting
        production of 58
Rocketry, history of early 5
Rock formations
    microwave measurement of 46
    radar interpretation and mapping of
        91
    remote sensing applied to 146
    See also Geology
Rocky Mountains, space, air, and
        photo images of 251
Rural planning, bibliography on 290
Russia. See Union of Soviet Socialist
        Republics

S

Salinity studies, microwave measure-
        ment of 68
Sand
    microwave measurement of 46
    Skylab observations of 116
    survey of using spectral
        luminance 86
    See also Deserts
San Francisco Bay Region, satellite
        maps of 282
Saskatchewan, Canada, remote
        sensing of water supply in
        153
Satellites
    surveillance techniques of 148
    types of and treatment of data
        from 78
    use of in development of a remote
        sensing global information
        system 65
    See also Spacecraft; names of
        satellites and satellite
        systems (e.g. Skylab)
Satellites, meteorological 66
    analysis and interpretation of data
        from 9
    description of programs conducted
        by 135A
    See also Atmosphere; Nimbus
        (satellite); Weather and
        climate
SEASAT (satellite), economic assess-
        ment of 25
Self-scanned solid state sensors 195
Shipping
    application of SEASAT to 25
    monitoring of by radar 96
Side Looking Airborne Radar. See
        Radar
Skylab 236
    application of to European earth
        resources study 147
    catalog of volcano photographs
        from 246
    imagery catalogs and photo
        indexes from 250, 257-58,
        267

operational aspects of and analysis
    of data from 140, 184
outline and preview of the project
    5, 12, 110, 116, 124
photographs of the U.S. based on
    33
    of the Rocky Mountains 251
See also Earth Resources Experiment
    Package
Snow
    bibliography on 283a
    mapping and observation of 116,
        182
    microwave measurement of 46, 183
Soils
    aerial mapping of 127, 135, 166
    bibliography on mapping of 294B
    microwave applications to 30, 46,
        68, 70, 183
    radar interpretation and mapping
        of 91
    remote sensing applied to 146, 153
Sonar
    in ice studies 165
    introduction to the techniques of
        48
Sound wave physics 8
South Africa, Republic of, application
    of ERTS data to 162
South Dakota School of Mines and
    Technology, remote sensing
    workshops of 377
Spacecraft
    description and illustration of 18
    photographic methods employed on
        14
    See also Satellites; names of space
        flights (e.g. Gemini space
        flight)
Spacelab 9
Space photography 14, 34, 114, 148
    bibliography on 314
    in earth resource monitoring 19,
        93
    equipment and techniques of 211
    for geography 31
    history of 72
    journals about 333
    representative examples of 20,
        73-74
        index to 262

in urban and regional planning
    43
See also Aerial photography
Space program
    benefits of 192, 200
    impact of 160A
    overview of the first 15 years of
        157
Space shuttle and transportation 65
    operation and objectives of 112,
        117
    payloads and utilization of 178
    radar applications in 96
    role of in the earth resources
        program 30
Spanish America. See Latin America
Spectral luminance, use of to measure
    sand deposits 86
Spectrometer, in city planning
    studies 16
Spectroscopy. See Infrared spectro-
    scopy; Raman spectroscopy
Stanford University, remote sensing
    workshops of 378
State government
    activities of in remote sensing 4
    catalogs of remote sensing imagery
        available from 250A
    satellite applications by and legis-
        lations of 10
    space program imagery radar and
        96
Sun, photographic selections of the
    73
Surveying, earth, remote sensing
    applied to 149
Surveyor, photographs from 20
Swamps, marine, microwave measure-
    ment of 46
Synthetic Aperature Radar. See
    Radar

T

Television
    in aerospace surveillance 148
    in remote sensing 177
        course notes on 223
Tennessee Valley Authority. See
    U.S. Tennessee Valley
    Authority

Thailand, ERTS application to rice production in 58
Thematic mapper 103
Thermal imagery 217
   in geomorphology 126A
   in soil mapping 135
Thermal surveys, use of aircraft for 29
Tidal inlets, catalog of aerial photographs of 240. See also Coastal waters
Tiros satellites, earth photographs from 241. See also Improved TIROS Operation System
Topography, radar interpretation and mapping of 91. See also Geography
Total Earth Resources System for the Shuttle Era 30
Traffic control. See Transportation planning
Transportation planning
   air and space photography in 43, 294C., 297, 304
   bibliography on 292, 294C, 297, 304
   space technology applied to 66, 158
Transportation systems, atlases showing 276
Tropical rain forests, aerial photography in assessment of the volume of 166

U

Ultraviolet waves, physics of 8
Underdeveloped countries. See Developing countries
Union of Soviet Socialist Republics
   current and future earth observation programs of 179A
   space photographs from the 20
United Nations
   role of in international cooperation for space exploration 192
   role of in remote sensing 174
United States
   current and future earth observation programs of 179A

maps and photographs of 33, 205, 273, 283
U.S. Air Force, history of reconnaissance programs of 39A
U.S. Army. Corps of Engineers
   bibliography of reports of 315A
   catalog of remote sensing imagery available from 250A
U.S. Army. Corps of Engineers. Coastal Engineering Research Center, Coastal Imagery Data Bank 256
U.S. Army, remote sensor capability study of 84
U.S. Geological Service, bibliography of reports of 287
U.S. National Aeronautics and Space Administration
   application of space technology to peaceful uses by 62
   bibliography of technical letters of 287
   history of 5
   index to photographs available from 262
   origin of names used by 131
   See also names of programs (e.g. Landsat Follow-On Program)
U.S. National Aeronautics and Space Administration. Goddard Space Center, the ERTS program and 13
U.S. National Aeronautics and Space Administration. Manned Spacecraft Center, activities of 115
U.S. National Environmental Satellite Service, catalog of operational satellite products of 243, 247
U.S. National Oceanographic and Atmospheric Administration
   cloud data available from 270
   snow observation activities of 182
U.S. Tennessee Valley Authority, bibliography of Materials of use to 306
Universe, atlas of 272
Urban planning 16, 27

air and space photography in 16, 43, 166, 255, 294C, 304, 307

bibliographies on 284, 288, 290, 294, 294C, 297, 304, 307, 312A, 314, 317

radar interpretation and mapping in 91

See also Transportation planning

V

Vegetation
aerial mapping of 91, 127
aerial photography of damage to 166
application of remote sensing and space technology to 118, 146–47
bibliography on 298
microwave measurement of 70, 183
radar analysis of 96
Skylab observations of 116
See also Agriculture
Vegetation, aquatic, mapping of 189
Vertebrate populations, remote sensing of 149
Volcanoes, catalog of satellite photographs of 246
Voskhod 2 (manned spacecraft), optical investigations from 149

W

Washington, D.C., maps of land use in 278
Water pollution. See Pollution

Water resource planning
aerial photography and 166
application of remote sensing and space technology to 17A, 65, 122, 146, 155, 196
bibliographies on 305, 309, 314
ERTS program and 38, 58, 136, 149, 160, 235
expected benefits of Landsat to 23
journals about 367
microwave applications in 96, 183
See also Floods; Groundwater; Hydrography; Hydrology; Irrigation; Oceanography; Wetlands
Water resources, atlases showing 276
Watershed modeling 153
Watson-Watt, Robert 129
Waves. See Oceanography
Weather and climate
application of remote sensing and space technology to 65, 146, 149, 201
journals about 350, 352
meteorological records available on 271
See also Atmosphere; Clouds; Meteorological phenomena; Satellites, meteorological
Wetlands, aerial photography in the monitoring of 166
Wind, microwave measurement of 70
World Administrative Radio Conference (1979) 70

NTIS INDEX

Many of the publications which are included in this resource guide are reports and publications of the U.S. National Aeronautics and Space Administration and other U.S. government agencies. In addition, many have been sponsored and funded by various functions of the U.S. government. Such publications are generally indexed by the National Technical Information Service (Spring-field, Virginia) and are listed in various indexes published by NASA, including SCIENTIFIC AND TECHNICAL AEROSPACE REPORTS (STAR) (citation #364) and EARTH RESOURCES: A CONTINUING BIBLIOGRAPHY WITH INDEXES (citation #313). Each of the citations in these two publications is identified by an accession number, and often in the open literature, this number is pro-vided as an aid for securing the publications references. This NTIS Index pro-vides a listing of these numbers, in sequence, of those publications which are included in this resource guide. As such, it provides a cross reference for those particular citations.

AD A049 351	315A	
AD 465 778	127	
AD 648 818	47	
AD 721 653	226 ·	
AD 723 061	303	
AD 751 192	128	
AD 755 508	256	
AD 766 720	310	
N 65 29925	127	
N 65 30350	156	
N 65 33550	176	
N 66 37028	62	
N 66 37029	62	
N 66 37030	62	
N 67 13461	176	
N 67 23338	122	
N 67 24785	46	
N 67 24793	46	
N 67 26907	47	
N 68 19870	296	

N 68 33392	148
N 68 36402	293
N 69 22267	144
N 69 27754	66
N 69 27755	66
N 69 27876	66
N 69 27962	66
N 69 27963	66
N 69 28072	66
N 69 28102	66
N 69 28160	66
N 69 28240	66
N 69 28360	66
N 69 28505	300
N 69 28938	66
N 69 30303	316
N 69 33626	176
N 69 35000	101
N 69 38951	198
N 70 14072	151

N 70 14446	77
N 70 14447	77
N 70 14448	77
N 70 34409	104
N 70 34410	104
N 70 34411	104
N 70 34412	104
N 70 34413	104
N 70 34414	104
N 70 34415	104
N 70 34416	104
N 70 34417	104
N 70 34418	104
N 70 34419	104
N 70 34420	104
N 70 34421	104
N 70 34422	104
N 70 34423	104
N 70 34424	104
N 70 34425	104
N 70 34426	104
N 70 34427	104
N 70 34428	104
N 70 38529	283A
N 70 42766	312
N 71 12199	268
N 71 12566	73
N 71 14697	283A
N 71 25256	198
N 71 26398	312
N 71 35479	291
N 72 13302	269
N 72 13303	269
N 72 13851	236
N 72 16085	173
N 72 16382	194
N 72 18324	121
N 72 18331	19
N 72 20958	107
N 72 21357	308A
N 72 23307	107
N 72 23324	198
N 72 24423	306
N 72 26272	197
N 72 26285	197
N 72 26811	109
N 72 30319	187
N 72 31382	186
N 73 12153	26
N 73 12401	26
N 73 12418	26
N 73 12419	26
N 73 11409	71
N 73 13829	198
N 73 13829	200
N 73 15198	26
N 73 15200	26
N 73 16348	194
N 73 22368	256
N 73 22403	214
N 73 22404	214
N 73 28207	159
N 73 28389	159
N 73 28405	159
N 73 28450	214
N 73 30317	169
N 73 30356	214
N 73 30370	176
N 73 30383	309
N 73 31867	198
N 73 32299	214
N 73 32301	214
N 74 10385	214
N 74 10549	160
N 74 10813	209
N 74 11161	160
N 74 11287	195
N 74 12202	310
N 74 16061	258
N 74 18055	17
N 74 18056	17
N 74 18057	17
N 74 18058	17
N 74 18059	17
N 74 18060	176
N 74 25898	247
N 74 27825	18
N 74 30705	160
N 74 30774	160
N 74 32795	235
N 74 32796	235
N 74 32797	235
N 74 32798	235
N 74 32799	235
N 74 32804	250
N 74 33873	160
N 75 12416	106
N 75 13010	108
N 75 14203	38
N 75 14204	38

N 75 14205	38
N 75 14206	38
N 75 14207	38
N 75 14208	38
N 75 14209	38
N 75 14210	38
N 75 14211	38
N 75 14212	38
N 75 14213	38
N 75 14214	38
N 75 14215	38
N 75 14216	147
N 75 15138	132
N 75 16050	155
N 75 16404	58
N 75 16405	58
N 75 16422	198
N 75 18705	64
N 75 20789	13
N 75 20798	267
N 75 20813	23
N 75 20814	23
N 75 20815	23
N 75 20816	23
N 75 20817	23
N 75 20818	23
N 75 20819	23
N 75 20820	23
N 75 20821	23
N 75 20822	23
N 75 20823	23
N 75 20824	23
N 75 20825	23
N 75 26477	220
N 75 27536	257
N 75 28502	244A
N 75 30634	210
N 75 31544	30
N 75 31545	30
N 75 31546	30
N 75 31547	30
N 75 31548	30
N 75 31549	30
N 75 31550	30
N 75 31551	30
N 75 31552	30
N 76 10556	245
N 76 10934	198
N 76 11811	175
N 76 13714	240
N 76 14583	238
N 76 14585	232
N 76 17469	196
N 76 17501	196
N 76 17552	196
N 76 17588	196
N 76 17613	196
N 76 18162	65
N 76 18195	65
N 76 18199	65
N 76 18301	65
N 76 18623	65
N 76 18624	65
N 76 18625	65
N 76 18626	65
N 76 18629	49
N 76 18696	65
N 76 18749	65
N 76 18769	65
N 76 18988	66
N 76 18995	65
N 76 19529	31
N 76 19537	103
N 76 19538	103
N 76 20813	23
N 76 20814	23
N 76 20815	23
N 76 20816	23
N 76 20817	23
N 76 20818	23
N 76 20819	23
N 76 20820	23
N 76 20821	23
N 76 20822	23
N 76 20823	23
N 76 20824	23
N 76 20825	23
N 76 23664	4
N 76 23673	285
N 76 26631	196
N 76 26646	196
N 76 27644	34
N 76 28614	25
N 76 28615	25
N 76 28616	25
N 76 28617	25
N 76 28618	25
N 76 28619	25
N 76 28620	25
N 76 28621	25

N	76 28622	25
N	76 28623	25
N	76 29686	30
N	76 29687	30
N	76 29693	289
N	76 30635	219
N	76 33595	22
N	76 33596	22
N	77 10619	10
N	77 14540	150
N	77 18309	96
N	77 18536	246
N	77 21531	308
N	77 25615	84
N	77 27472	244
N	78 10509	312A
N	78 14464	176
N	78 14529	176
N	78 17436	244B
N	78 17437	244B
N	78 17438	244B
N	78 17439	244B
N	78 19584	315A
N	78 23509	17A
N	78 28588	250A
N	78 28593	3A
N	79 10497	68
N	79 10505	294A
N	79 10506	294C
N	79 10507	294B
PB	192 863	301
PB	194 072	283A
PB	195 748	302
PB	199 123	307
PB	202 726	291
PB	211 101	317
PB	214 449	43
PB	220 154	1
PB	224 754	102
PB	242 813	238
PB	243 933	232
PB	248 294	285
PB	264 171	69
PB	283 027	312A
TT	64 11094	52

SERIES INDEX

Because many publishers of the material contained in this resource guide have published a wide variety of materials, they often find it convenient to present works dealing with a single subject or set of similar subjects under a "series" title. Also, because some researchers are familiar with the series of the several publishers, this index is provided to aid them in finding remote sensing publications contained in the several series. This index also provides a cross reference for publications of a single publisher within a single series which are referenced herein. It is alphabetized letter by letter, and numbers refer to entry numbers.

A

Academic Press--Electrical Sciences:
   A Series of Monographs and
   Texts 36
Advisory Group for Aerospace Research
   and Development
   Conference Proceedings No. 29 148
   Conference Proceedings No. 90 173
American Astronautical Society
   Science and Technology Series v.
      23 158
   Science and Technology Series v.
      30 178
   Science and Technology Series v.
      31 157
   Advances in Astronautical Sciences
      v. 26 179
   Advances in Astronautical Sciences
      v. 31 184
American Geographical Society
   Special Publication No. 4 53
American Water Resources Association
   Proceedings No. 17 189

C

Council of Planning Librarians

Exchange Bibliography Series
   No. 69 294
Exchange Bibliography Series
   No. 116 297
Exchange Bibliography Series
   No. 222 284
Exchange Bibliography Series
   No. 352 284
Crowell Company Pictorial Guides
   74

D

Doubleday and Company
   Science Study Series No. S-26
      75
   Science Study Series No. S-38
      76
Dowden, Hutchinson and Ross
   Community Development Series
      130
Duxbury Press
   The Man-Environment System in
      the Late Twentieth Century 88

Series Index

E

Eastman Kodak Company
    Publication M-28 24
    Publication M-76 204
European Space Research Organization
    Special Publication (ESRO SP) 100
    147

F

Federation of Rocky Mountain States
    Technical Report No. 28 251

H

Harvard (University)
    City Planning Studies 16
Hemisphere Publishing Company
    Advances in Thermal Engineering
    5 199

I

Iowa Geological Survey
    Public Information Circulars No. 3
    167

J

Joint Publication Research Service
    Publication (JPRS) 58739 169

K

Kansas University Remote Sensing
    Laboratory
    Technical Report RSL-100 295

L

Laboratory of Applied Remote Sensing
    Information Note No. 110474 220
    Information Note No. 052576 219
Los Alamos Scientific Laboratory--
    Reports LA-6279-MS 246
Louisiana State University--School of
    Geoscience Miscellaneous Publi-
    cations No. 72 288

M

McGraw-Hill Civil Engineering
    Series 60
McGraw-Hill Handbook Series 98
McGraw-Hill International Series in
    the Earth Sciences 224
Macmillan--Air Force Academy Series
    39A
Macmillan Monographs in Applied
    Optics 41

P

Prentice-Hall--Series in Electronic
    Technology 134

T

Texas A & M University Remote
    Sensing Center
    Technical Report RSC-22 308A

U

UNESCO--Natural Resources Research
    v. 6 191
U.S. Department of Agriculture--
    Economic Research Service
    Agricultural Handbook No. 372
    205
    Agricultural Handbook No. 384
    205
    Agricultural Handbook No. 406
    205
    Agricultural Handbook No. 409
    205
    Agricultural Handbook No. 419
    205
U.S. Department of Agriculture
    Technical Bulletin No. 1344 82
U.S. Department of Army--Coastal
    Engineering Research Center
    Miscellaneous Papers No. 2-73
    310
    Miscellaneous Papers No. 3-72
    256
U.S. Department of Army--Natick
    Laboratories
    Technical Reports No. 71-27-ES
    (Series ES-61) 303
U.S. Geological Survey--Circular
    693 18

U.S. Geological Survey
  Interagency Reports No. 129   293
  Interagency Reports No. 189   301
  Interagency Reports No. 203   302
  Interagency Reports No. 204   307
  Interagency Reports No. 234   306
  Interagency Reports No. 250   43
  Interagency Reports No. 253   238
U.S. Geological Survey
  Professional Papers No. 373   80
  Professional Papers No. 590   242
  Professional Papers No. 929   136
  Professional Papers No. 1015   202
U.S. Geological Survey
  Technical Letters--NASA No. 86
    287
  Technical Letters--NASA No. 131
    304
  Technical Letters--NASA No. 134
    300
U.S. Geological Survey--Miscellaneous
  Investigation Series
  Map I-858   278
U.S. Geological Survey--Open File
  Reports No. 75-196   232, 282,
    287
U.S. National Aeronautics and Space
  Administration
  Educational Publications EP-103   233
  Educational Publications EP-111   124
  Facts NF-43/1-72   111
  Facts NF-44/7-72   112
  Facts NF-54/1-75   113
  Facts NF-56/1-75   110
  Facts NF-57/1-75   114
  History Series   131
  SP (Special Publications) No. 129
    119
  SP (Special Publications) No. 142
    122
  SP (Special Publications) No. 168
    20
  SP (Special Publications) No. 171
    118
  SP (Special Publications) No. 195
    198
  SP (Special Publications) No. 200
    50
  SP (Special Publications) No. 206
    15
  SP (Special Publications) No. 230
    120

  SP (Special Publications) No. 250
    73
  SP (Special Publications) No. 251
    198
  SP (Special Publications) No. 275
    19
  SP (Special Publications) No. 283
    194
  SP (Special Publications) No. 285
    121
  SP (Special Publications) No. 286
    198
  SP (Special Publications) No. 294
    197
  SP (Special Publications) No. 312
    198
  SP (Special Publications) No. 313
    200
  SP (Special Publications) No. 327
    159
  SP (Special Publications) No. 331
    198
  SP (Special Publications) No. 335
    195
  SP (Special Publications) No. 351
    160
  SP (Special Publications) No. 356
    160
  SP (Special Publications) No. 360
    95
  SP (Special Publications) No. 361
    198
  SP (Special Publications) No. 364
    155
  SP (Special Publications) No. 370
    170
  SP (Special Publications) No. 376
    175
  SP (Special Publications) No. 380
    116
  SP (Special Publications) No. 384
    198
  SP (Special Publications) No. 391
    182
  SP (Special Publications) No. 400
    12
  SP (Special Publications) No. 407
    117
  SP (Special Publications) No. 410
    275

SP (Special Publications) No. 4402
131
SP (Special Publications) No. 4403
5
SP (Special Publications) No. 5099
211
SP (Special Publications) No. 7036
314
SP (Special Publications) No. 7036
(01) 315
SP (Special Publications) No. 7041
313, 331, 366
SP (Special Publications) No. 7050
123
Technical Translations F-681 71
Technical Translations F-16852 31
Technical Translations F-16924 34
U.S. National Environmental Satellite
Service Key to Meteorological
Records
Documentation Series No. 5.3 271
Documentation Series No. 5.4 270

U.S. National Oceanic and Atmos-
pheric Administration National
Environmental Satellite Service
Technical Memorandum (NESS) No.
53 247
Technical Memorandum (NESS) No.
88 243
Technical Memorandum (NFSCTM)
No. 10 283A

W

Water Resources Scientific Information
Center
Bibliography Series No. 73-211
309
Wiley--Series on Photographic Science
and Technology in the Graphic
Arts 230
Wiley-Interscience: Environmental
Science and Technology Series
39